Communications in Computer and Information Science 671

Commenced Publication in 2007
Founding and Former Series Editors:
Alfredo Cuzzocrea, Dominik Ślęzak, and Xiaokang Yang

More information about this series at http://www.springer.com/series/7899

Seok-Won Lee · Takako Nakatani (Eds.)

Requirements Engineering Toward Sustainable World

Third Asia-Pacific Symposium, APRES 2016 Nagoya, Japan, November 10–12, 2016 Proceedings

Springer

Editors
Seok-Won Lee
Department of Software and Computer Engineering
Ajou University
Suwon, Kyonggi-do
Korea (Republic of)

Takako Nakatani
The Open University of Japan
Chiba
Japan

ISSN 1865-0929 ISSN 1865-0937 (electronic)
Communications in Computer and Information Science
ISBN 978-981-10-3255-4 ISBN 978-981-10-3256-1 (eBook)
DOI 10.1007/978-981-10-3256-1

Library of Congress Control Number: 2016957848

Printed on acid-free paper

This Springer imprint is published by Springer Nature
The registered company is Springer Nature Singapore Pte Ltd.
The registered company address is: 152 Beach Road, #22-06/08 Gateway East, Singapore 189721, Singapore

Preface

APRES 2016 was the third edition of the Asia-Pacific Requirements Engineering Symposium (APRES 2016) that serves as a highly interactive forum for in-depth discussion of all issues related to requirements engineering following the success of APRES 2014 in New Zealand and APRES 2015 in China.

The symposium took place on the campus of Nanzan University, in the historical city of Nagoya, which is known as Nagoya castle, and the heart of Japanese quality engineering, including one of the world's largest community of the automotive industry.

Requirements engineering (RE) as an established discipline of research and practice in software and systems development has a reputable RE community worldwide. The importance of developing and following effective RE practices has long been recognized by researchers and practitioners alike, especially when facing the latest advancements in sociotechnological needs in various industrial sectors, such as e-commerce, manufacturing, health care, etc.

This year, the discussions concentrated on "sustainable world." Sustainability is one of the emerging quality attributes in complex software systems that greatly affects our daily lives in many aspects, and in varying contexts is not supported by traditional software engineering methods. We address the needs of supporting sustainability in RE to have a substantial impact on our society.

According to the growing interest in RE research and practice, the APRES symposium series aims to develop and expand the RE research and practice community specifically in the Asia-Pacific region and to foster collaborations among local researchers and practitioners in Asia and Oceania.

This year we received 14 submissions by active researchers from all over the world, among which seven full papers and three short papers were accepted. All papers were carefully reviewed by at least three Program Committee members, and detailed constructive feedback was provided to the authors. We had participants and presentations from Asia, Oceania, and North America on RE topics, such as requirements traceability and prioritization, requirements modeling and process for quality, requirements validation, and requirements analysis.

APRES 2016 had a focused program including three keynote speeches covering RE research and big industry challenges. There were three paper presentation sessions, including RE methodologies and qualities. A plenary panel was arranged at the end of the conference to discuss future steps for RE improvement such as: automotive RE, artificial intelligence in RE, along with industry panels. APRES 2016 also attracted a good number of participants and enriched the overall offering of the conference for the discussion of "real-world" problems and sharing of good industry practices in RE, especially so as to foster RE research–industrial collaborations in the Asia-Pacific region.

Our greatest thanks go to the authors and presenters, whose contributions made APRES 2016 a success. We are grateful to the Program Committee members for their thorough and timely reviews on the submissions. We thank the Steering Committee for their valuable guidance. We would like to thank Prof. Mikio Aoyama and Prof. Naoyasu Ubayashi, the general co-chairs, for organizing and managing the industrial track of APRES 2016. We thank the sponsors of APRES 2016, Nanzan University, and many other industry partners.

Our thanks also go to Springer, the publisher of the APRES proceedings, for their continuous support. Finally, thanks to EasyChair for making conference management such a straightforward task.

We hope you all enjoy the APRES 2016 proceedings.

November 2016

Seok-Won Lee
Takako Nakatani

Organization

General Chairs

Mikio Aoyama	Nanzan University, Japan
Naoyasu Ubayashi	Kyushu University, Japan

Program Chairs

Seok-Won Lee	Ajou University, South Korea
Takako Nakatani	The Open University of Japan, Japan

Publicity Chair

Mari Inoki	Kogakuin University, Japan

Organizing Committee Chairs

Masami Noro	Nanzan University, Japan
Atsushi Sawada	Nanzan University, Japan
Atsushi Yoshida	Nanzan University, Japan

Secretary and Web Master

Sayuri Taki	Nanzan University, Japan

Program Committee

Mikio Aoyama	Nanzan University, Japan
Muneera Bano	University of Technology Sydney, Australia
Jim Buchan	Auckland University of Technology, New Zealand
Xiaohong Chen	Eastern China Normal University, China
Tony Clear	Auckland University of Technology, New Zealand
Smita Ghaisa	Tata Research Design and Development Center, India
Naveed Ikram	Riphah International University, Pakistan
Mari Inoki	Kogakuin University, Japan
Zhi Jin	Peking University, China
Massila Kamalrudin	Innovative Software System and Services Group, Malaysia
In-Young Ko	Korea Advanced Institute of Science and Technology, South Korea
Seok-Won Lee	Ajou University, South Korea

Zhi Li	Guangxi Normal University, China
Lin Liu	Tsinghua University, China
Xiaodong Liu	Edinburgh Napier University, UK
Stuart Marshall	Victoria University of Wellington, New Zealand
Saeko Matsuura	Shibaura Institute of Technology, Japan
Pornsiri Muenchaisri	Chulalongkorn University, Thailand
Hiroyuki Nakagawa	Osaka University, Japan
Takako Nakatani	The Open University of Japan, Japan
Soojin Park	Sogang University, South Korea
Rong Peng	Wuhan University, China
Shinobu Saito	Nippon Telegraph and Telephone Corporation, Japan
Shahida Sulaiman	Universiti Teknologi Malaysia, Malaysia
Yijian Wu	Fudan University, China
Yong Xia	IBM, China
Rieko Yamamoto	Fujitsu Laboratories Ltd., Japan
Hongji Yang	Bath Spa University, UK
Junbeom Yoo	Konkuk University, South Korea
Eric Yu	University of Toronto, Canada
Mansooreh Zahedi	IT University of Copenhagen, Denmark
Haiyan Zhao	Peking University, China
Didar Zowghi	University of Technology, Sydney, Australia

Steering Committee

Mikio Aoyama	Nanzan University, Japan
Jim Buchuan	Auckland University of Technology, New Zealand
Zhi Jin	Peking University, China
Lin Liu	Tsinghua University, China
Rong Peng	Wuhan University, China
Didar Zowghi	University of Technology, Sydney, Australia

Sponsors

Academic Sponsor

Information Processing Society of Japan

Host and Platinum Sponsors

Nanzan University

DENSO CREATE INC

NIPPON TELEGRAPH AND TELEPHONE CORPORATION

Gold Sponsors

FUJITSU LIMITED

ITOCHU Techno-Solutions Corporation (CTC)

JISA (Japan Information Technology Services Industry Association)

Silver Sponsor

QUNIE CORPORATION

Keynote Papers

In Requirements Engineering, All Roads Lead to ROME

Carl K. Chang

Department of Computer Science, Iowa State University,
Ames, IA 50011, USA

Abstract. As shown in the title of this talk, ROME carries two meanings. For a requirements engineer, ROME stands for "Requirements Of Meeting Expectations". On the other hands, for end users, ROME stands "Requirements Of Matching Experiences". In the global requirements engineering (RE) enterprise, both meanings had produced profound impact to varying stakeholders when a modern computer system is envisioned, analyzed, designed, implemented, deployed and eventually entered into the operation and maintenance mode.

In this talk, I will focus on an emerging field named situation analytics (SA) [1], and explain how it supports a broad spectrum of computer system development, maintenance and evolution activities in the emergent era of Internet of Things.

I will review some of the defining concepts for SA based on my earlier work in Situ [2]. When we tailor SA to RE, I will elaborate on my recommendation that all roads (i.e. activities in our profession) lead to ROME.

References

1. Chang, C.K.: Situation Analytics – a foundation for a new software engineering paradigm. An Outlook Article in Comput. **49**(1), 24–33 (2016)
2. Chang, C.K., et al.: Situ: a situation-theoretic approach to context-aware service evolution. IEEE Trans. Serv. Comput. **2**(3), 261–275 (2009)

Vehicle New Capabilities as a Social Infrastructure

Yui Inoue

Chairman of the Board, TOYOTA InfoTechnology Center Co., Ltd., 6-6-20 Akasaka, Minato-ku, Tokyo, 107-0052, Japan

Abstract. ITS, Intelligent Transport System, with DSRC, Dedicated Short Range Communications, was launched in Japan last year as the first DSRC system in the world. Vehicle2vehicle communications became available among some models of new vehicles with DSRC to avoid such as inter-section accidents and rear-end collisions. Vehicle2Infrastructure (traffic signals) was also introduced last year and will be expanded gradually to assist, for an example, handicapped pedestrians by extending traffic light-length.

Automated driving and/or ADAS, Advanced Driving Assist System, will further renovate social infrastructures by its powerful wireless communications and information processing capability. Its capability will not be limited only to mobility aspects but to new social activities. One of the first deployment will be for providing resilient communications systems for residences and public agencies when a huge deserter destroys existing telecom based communications in wide areas. Once this capability is installed in vehicles, its usage will not be limited to emergency cases only but to ordinal activities in the society to complement existing information and communication systems.

Another example of them will be IoT, Internet of Things, based new society and industry supported by vehicles. IoT agriculture, fishery, forestry, and mining will be enabled to increase production quantity and to improve quality and traceability by the vehicle new capabilities which will be quite popular and close to production in their fields.

Strategy Rules
Five Timeless Lessons from Bill Gates, Andy Grove, and Steve Jobs

Michael A. Cusumano

Sloan Management Review Distinguished Professor of Management,
Sloan School of Management, Massachusetts Institute of Technology,
Cambridge, MA 02142, USA

Abstract. As This talk focuses on a new book by David Yoffie of Harvard Business School and Michael Cusumano of MIT, titled Strategy Rules: Five Timeless Lessons from Bill Gates, Andy Grove, and Steve Jobs (Harper Business, 2015. Now translated into 17 languages, including Japanese translation in 2016).

The founders of Microsoft, Intel, and Apple had very different backgrounds and personalities, but approached strategy in remarkably similar ways. The talk describes the five rules or principles they followed, based on close observation of the three CEOs by the authors for the past 25 years.

The first rule is "Look Forward, Reason Back." Gates, Grove, and Jobs all created very different companies but extrapolated from the same Moore's Law. They first determined where they wanted their companies to be at a given point in the future, and then "reasoned back" to identify the moves that would take them there.

The second is "Make Big Bets, Without Betting the Company." All three made enormous strategic bets to grow and shape their markets, but rarely took gambles that put the financial viability of the company at undue risk.

Third is "Build Platforms and Ecosystems." Technology leaders often need to think beyond the boundaries of their own firms to harness a large user base and network effects. The goal should be to create "industry platforms" rather than standalone products that enable other firms to create complementary products and services that make the platforms increasingly valuable.

Fourth is "Exploit Leverage and Power: Play Judo and Sumo." These firms were clever, quick, and nimble, especially in their early years, and they all used market power when they had it.

The final rule is "Shape the Company around Your "Personal Anchor." All three CEOs learned how to compensate for their weaknesses as they built companies around their personal strengths – Gates' deep understanding of early PC software, Grove's devotion to engineering discipline, and Jobs' obsession with designs and user interfaces that made technology accessible to the average person.

The Future of Requirements Engineering

Sensing Technologies Required for ADAS/AD Applications

Kazuoki Matsugatani

Director, ADAS Business and Technology Development Div.,
DENSO CORPORATION. 1-1 Shouwa-cho, Kariya, 448-8661, Japan

Abstract. ADAS (Advanced Driver Assistance System) and AD (Automated Driving) have received much attention in recent years. Various firms and companies are developing these systems actively.

Vehicle driving task consists of three functions; perception, decision and control. For ADAS, a portion of perception and some control are automated. And for AD, all of functions including decision are automated. Throughout these applications, perception plays an important role, and supporting perception by sensing devices makes driving much safer.

In this talk, firstly, I focus on surround observation sensors used for perception. Typical sensors, camera, radar and LIDAR (Light Detection and Ranging) are introduced and their functions are explained. These components designed for detecting various objects to maintain safe driving.

Then I introduce wireless communication and HMI (Human Machine Interface) devices. Wireless device connects a vehicle to infrastructure, and HMI connects vehicle to human driver. For wireless communication, local or ad-Hoc media of vehicle to infrastructure and inter vehicle communication are utilized. And cellular based radio is also used for vehicles. Regarding HMI devices, DSM (Driver Status Monitor) and HUD (Head Up Display) are typical components required for ADAS/AD.

Finally, I show our recent activities for investing and demonstrating ADAS/AD applications. Our ultimate objective is 'ZERO ACCIDENTS', and to make driving 'FUN'. We have been challenging persistently to realize this objective.

Machine Learning as a Programming Paradigm and its Implications to Requirements Engineering

Hiroshi Maruyama

Chief Strategy Officer, Preferred Networks, Inc., Otemachi Bldg. 2F, 1-6-1, Otemachi, Chiyoda-ku, Tokyo, 100-0004, Japan

Abstract. Application areas of machine learning is quickly expanding, from image recognition to language translation, game playing, and autonomous driving. Machine learning can be viewed as a tool for building a system inductively from a set of input-output examples, where specifications of such a system are given as training data sets. How should we translate our requirements into these specifications, that is, how should we prepare appropriate training data sets that reasonably represent the given requirements? Since new training data sets become continuously available for online systems, the specifications also continuously change over time. How should we assure all these specifications are consistent with the requirements?

There are a number of open questions with this new programming paradigm. This talk will explore some of these open questions and discuss their implications and the opportunities to the requirements engineering community.

REBOK and Requirements Engineering for the Digital Business Age

Mikio Aoyama

Department of Software Engineering, Nanzan University, 18 Yamazato, Showa-ku, Nagoya, 466-8673, Japan

Abstract. A digital transformation of the business and society is undergoing. It is driven by IT (Information Technology). It changes the way of business and the use of IT; from improving the business performance by the supportive use of IT, to the creation of new business value by the creation of digital business centered around IT. To make the digital business happen, requirements engineering is expected to play a key role to create innovative business models and the requirements.

Backed by JISA (Japan Information technology Services industry Association), a group of researchers and practitioners have been working together to create a knowledge platform of requirements engineering for more than a decade. As a product, REBOK (Requirements Engineering Body Of Knowledge) was published in 2011. It bridges from business requirements to systems requirements and software requirements. REBOK and a series of publications have been well received.

However, there are several similar but different disciplines and BOKs emerged in the business requirements arena, including BABOK (Business Analysis Body Of Knowledge), BIZBOK (Business Architecture Body of Knowledge), and CBOK (Business Process Management Common Body Of Knowledge). Furthermore, the disciplines of design thinking and related techniques are spreading. It is necessary to provide a comprehensive knowledge platform for the requirements engineering of digital business.

This talk overviews the current status and future direction of requirements engineering with respect to creating innovative business requirements for digital businesses.

References

1. Aoyama, M., Nakatani, T., Saito, S., Suzuki, M., Fujita, K., Nakazaki, H., Suzuki, R.: A model and architecture of REBOK(requirements engineering body of knowledge) and its evaluation. In: Proceedings of APSEC 2010, pp. 50–59. IEEE Computer Society, November 2010
2. Aoyama, M.: Bridging the requirements engineering and business analysis toward a unified knowledge framework. In: Proceedings of MReBA 2016, ER 2016 Workshops, LNCS 9975, Springer, November 2016, 10 pages (To Be Published)

Contents

Requirements Analysis

Requirements Traceability and Prioritization

Ensuring Traceability in Modeling Requirement Using Ontology Based Approach

Theresia Ratih Dewi Saputri[1] and Seok-Won Lee[2(✉)]

[1] Department of Computer Engineering, Ajou University, Suwon, Korea
trdsaputri@ajou.ac.kr
[2] Department of Software and Computer Engineering, Ajou University, Suwon, Korea
leesw@ajou.ac.kr

Abstract. Requirements traceability turns into essential principle in software engineering due to the needs to address evolving requirements in software system development. Requirements traceability helps to identify whether the entire requirements have been implemented consistently. However, the task to provide manual requirements traceability tends to become a costly and time-consuming procedure. It is unwise to invest the resources for manual monitoring and updating the traceability link in the requirements document. This work presents a solution to this problem by proposing an approach that uses an ontology-based knowledge representation along with information retrieval techniques. Ontology-based approach is used due to its ability to automatically generate the relationship among requirements concepts. In conclusion, the proposed approach is able to identify missing, broken or even new traceability links between the requirements artifacts.

Keywords: Requirements engineering · Traceability · Scenario-based approach · Ontology-based approach · Information retrieval

1 Introduction

The advancement of technology and the growth of user expectations lead to the changing environment in the software system development. In the consequence of changing environment, the developer usually faces the evolving requirements during the software deployment phase. The evolving requirements usually rise in the response to the changing of stakeholder needs or the missing part in the initial analysis.

These changes result in the need of reconfiguring, troubleshooting, and even deleting the requirements. Most often, they do not update their formal requirements document when there is a changing requirement in the development phase. This practice causes the inconsistency between textual requirements and the implementation [1]. The inconsistency enables the conflict among the requirements to arise. Hence, the change of the requirements document should be managed properly. Managing requirements is not only about managing the document, but also recognizing changes through continued requirements elicitation and evaluation of systems in their operational environment [2].

S.-W. Lee and T. Nakatani (Eds.): APRES 2016, CCIS 671, pp. 3–17, 2016.
DOI: 10.1007/978-981-10-3256-1_1

Therefore, one of the important aspects in defining the requirements of a system is imposing requirements traceability practice in the system development process. Requirements traceability can be defined as the ability to describe and to follow the life of the requirements [3]. By implementing requirements traceability, determining where the requirements come from, how the system was developed, and what are the effects of changing a certain requirement becomes easier. In order to improve requirements traceability, there are various approaches that have been proposed such as requirements traceability matrix. Most of them use their own traceability technique that is specific for their related approaches [4]. However, the task to provide manual requirements traceability tends to become a costly and time-consuming procedure. It is unwise to invest the resources for manual monitoring and updating the traceability link in the requirements document.

Concerning those problems, the proposed approach uses an ontology-based knowledge representation along with information retrieval techniques for extracting information from the scenario. In order to elicit the requirements, this research takes the advantages of the scenario-based approach. The gathered scenarios are used to extract requirements artifacts for constructing the knowledge base. This knowledge base is used to provide the traceability for the system. There are some challenges that should be faced when implementing scenario-based approach. Those challenges are (1) requirements knowledge is informal and (2) it is difficult to transform informal into formal model and to maintain it, due to its complexity [5]. Therefore, the aim of this work is to support requirements traceability with requirements knowledge acquisition from the scenario. This work also aims to identify the relations between sources of the requirements using ontology-based approach. Ontology is a widely known approach with a purpose to construct a conceptual domain modeling for knowledge engineering [5]. This ontology provides efficient searching and browsing materials to detect changes and to maintain requirements knowledge. By implementing an ontology-based approach for managing the evolving requirements, this research is able to automatically generate the relationship among requirements concepts. In order to evaluate our proposed approach, we implement the approach in the smart grid domain area. This field is chosen due to its various and complex stakeholder needed to be considered.

2 Background and Related Works

In this section, we briefly discuss the related work. Firstly, we explain the motivation behind requirements traceability. Secondly, we present one of the approach for implementing requirements traceability which is a scenario-based approach. Lastly, we discuss the use of ontology as knowledge representation in requirements traceability.

2.1 Requirements Traceability

The software system should be able to evolve in response to the environment in which these systems operate changes and stakeholder requirements change [2]. As the result, the requirements document should be maintained due to the growth of user expectations.

Therefore, one of the important methods in requirements engineering is requirements traceability.

Gotel and Finkelstein in [3] define requirement traceability as the ability to describe and follow the life of the requirements in both forwards and backwards direction as seen on Fig. 1. Pre-RS traceability depends on the ability to trace requirements from, and back to, their originating statement(s), through the process of requirements production and refinement, in which statements from diverse sources are eventually integrated into single requirements in the RS. On the other hand, post-RS trace-ability depends on the ability to trace requirements from, and back to, a baseline (the RS), through a succession of artifacts in which they are distributed. The changes to the baseline need to be re-propagated through this chain.

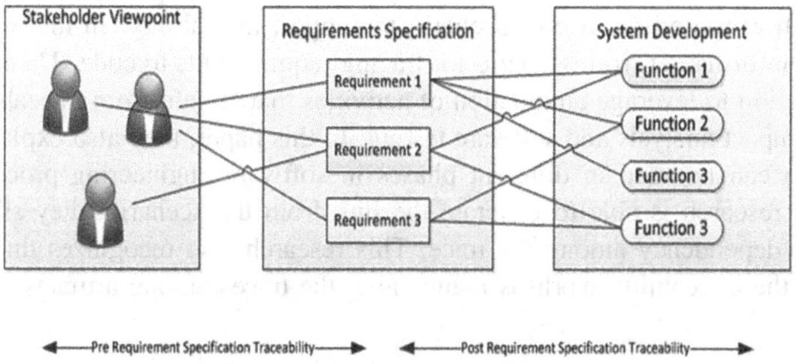

Fig. 1. Two basic type of requirements traceability. Pre requirements specification traceability traces the requirements from and to the original statement from stakeholders' viewpoints. Post requirements specification traceability trace the requirements from and to the software artifact.

By implementing requirements traceability, it will be easier to determine where the requirement comes from, how the system was developed, what is the effect of the changing the certain requirement and which requirement that is also affected by changing a certain requirement. Requirements traceability helps developers to control and manage the development and evolution of software. In [6], Heindl explains that traceability can help in managing the change and costs for all types of projects. Andrew Kannenberg et al. in [7] mentioned some of the importance of requirements traceability in software engineering process. By following traceability links, a project manager can quickly see how many artifacts will be affected by a proposed change and can make an informed decision about the costs and risks associated with that change.

There are some techniques that can be used to impose traceability in the software engineering process both manually and automatically such as template, requirements traceability matrices, and integrating documents. Davis A.M in [8] introduces the use of matrices for providing requirements traceability manually. Requirements traceability matrices contains of the many-to-many relationships among two baseline documents such as requirements document and test case document. This matrix can be used to analyze whether a certain requirement have been met with a particular test case. During the increase of complexity, most of researchers prefer the implementation of requirements traceability using automated tools [3].

2.2 Scenario-Based Approach

There are several methods that can be used to implement requirements traceability. One of the widely known methods is scenario-based approach. One of the advantages of scenario-based approach is the scenario can illustrate the need of stakeholders clearly. By using scenario, the analyst will be able to determine the precise interaction in the system based on the system description given by the stakeholders.

Egyed in [9] proposed scenario based approach to provide traceability in the requirements. A scenario makes the analyst salient to important events that should be taken into account in the uncertain future. That scenario based approach uses the determined trace information that has to be manually entered. Those trace information is gradually improved into trace links with the runtime information. However, this research does not support the automatically derived trace dependencies. Naslavsky et al. in [10] proposed the use of scenario to support traceability. In this work, they explore the information from scenario for tracing requirements to code. Then, they use that information to leverage automation of activities that benefit from traceability such as change impact analysis and software testing. In this paper, they also explained how the scenario can be used in different phases in software engineering process. Even though this research is able to generate the link from the scenario, they still cannot provide the dependency among the trace. This research also recognizes that a major problem in the traceability world is maintaining the traces among artifacts.

2.3 Information Retrieval Technique in Requirements Traceability

In order to overcome the difficulties to extract the information from scenario, some researches proposed the use of information retrieval technique [11]. Information retrieval is a technique to find material of an unstructured nature that satisfies an information need from within large collections [12]. The work by Luhn in [13] stated that information retrieval system could be designed based on comparison of content identifiers attached both the stored texts and to the user's information queries.

An information retrieval technique has been shown to assist with the automated generation of traceability links by reducing the time it takes to generate the traceability mapping [14]. The work proposed by Jane Cleland-Huang and Jin Guo in [15] shows how information retrieval technique can be used to create the trace link automatically. The research proposed the combination of information retrieval technique and ontology as a knowledge representation. In this paper, they also propose a classification scheme for categorizing the intelligence level of automated traceability techniques. This research also mention about the major limitation in requirements traceability area is the significant cost and effort needed to build an expert system. Therefore, our proposed approach also implements the information retrieval technique to extract the important term from scenario.

2.4 Ontology-Based Approach

Determining the relation among the requirements as the link becomes the important issue for building the requirements traceability tools for managing the evolving requirements. One of the major challenges in requirements traceability field is the missing traceability links among software artifacts. This problem result in the increasing number of efforts to provide the traceability links manually. Therefore, that problem becomes the main motivation to propose an automatic approach to generate the traceability links among software artifacts automatically. One of remarkable solution to address this problem is the used of ontology. Ontology is a conceptual modeling used to define the terms that are able to describe and represent a knowledge domain. Ontology is also known as formal specification for sharing and reusing includes hierarchy and a relation of concepts [16] which provides interoperability, browsing/searching, reuse, and structuring [17] for the requirements their engineering process.

Due to its capabilities, ontologies have been considered in several research areas such as information retrieval, knowledge management, electronic commerce and many other systems [18]. Moreover, the ontology concept allows constructing concise specifications of hierarchical conceptual structures and giving robust specifications of complex relationships among structural entities. In the area of requirements engineering, ontology is able to represent the structure of domain knowledge which reduce the ambiguity of requirements and handle the dynamic requirements. Zhang et al. in [19] proposed an ontology-based approach to for traceability recovery. In order to solve the problem, they create formal ontological representations for both the documentation and source code artifacts. By providing uniform ontological representations for various software artifacts, they are able to utilize semantic information conveyed by these artifacts and to establish their traceability links at semantic level. The work proposed by Aponte and Marcus in [27] proposed a method to determine traceability links among complex software artifacts. They built a traceability link recovery tool that implements different information retrieval techniques to generate ranked lists of candidate links.

Even though the previous studies performed remarkable work for using the ontology concept for traceability, the current ontology-based approaches miss the possibility to utilize the available textual information within artifacts. They only consider the structural information and rely on other mechanism for ontology population [20]. To address this issue, extracting and analyzing the requirements information from the scenario as the concepts in our ontology becomes main concern in our research.

3 Proposed Approach

As seen in Fig. 2, there are three main processes in our proposed approach which are Extracting Keyword, Defining Requirements Statements, and Mapping Requirements.

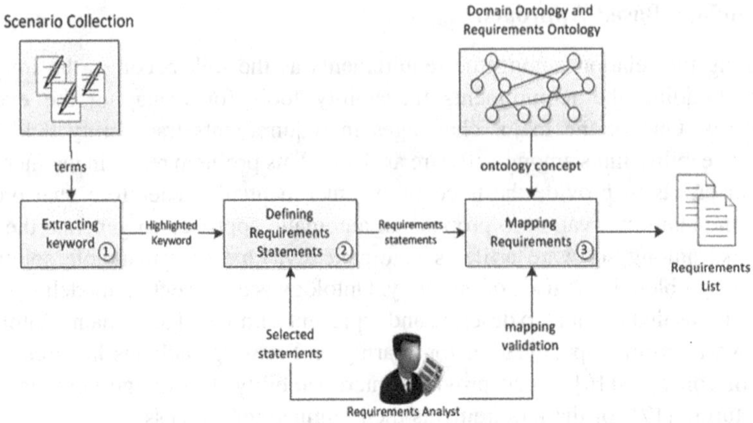

Fig. 2. Proposed approach

3.1 Extracting Keyword and Defining Requirements

The proposed approach uses scenario-based approach to elicit knowledge of the domain. A scenario provides not only clear functionalities for the system, but also the behaviors of the system. Moreover, the goal of stakeholder can be identified from a particular scenario. In the first process of the proposed approach, the keywords are extracted from the scenario. Those keywords are used to assist requirements analyst identify the requirements statements from the gathered scenario.

In order to extract the term from scenario, this work uses the Natural Language Processing (NLP) tool, Part-Of-Speech (POS) tagger called Stanford Post Tagger. POS Tagger is a piece of software that reads text in some language and assigns parts of speech to each word (and another token), such as noun, verb, adjective [21]. The work in [22] shows that the tagger is able to demonstrate the use of both preceding a following tag contexts via dependency network representation. It also shows the use of lexical features including jointly conditioning on multiple consecutive words.

This POS tagger provides tagged words and Named-entity Recognition (NER) to extract the terms from the text. The noun (NN, NNS, NNP) and verb (VB, VBG, VBN) indicate a stakeholder, an event or an action. The action and event are used to represent the software system behavior. Before extracting the terms of the text, the pre-processing raw data is needed, such as removing stop word, to reduce the noise of the data. In order to minimize the effort for analyzing the term, we highlight the extracted keywords based on the weight of each term. The weight is used to determine the domain relevance of these extracted terms. The second process in the proposed approach is defining the requirements statements. In this process, requirements analyst is asked to choose the requirements. In addition, to help requirements analyst, each candidate of the requirements statements is listed based on the highlighted keyword. Requirements analyst is able to select the requirements statements by clicking the keyword. Then, the sentence in which the clicked keyword comes from automatically becomes one of the selected requirements statements.

3.2 Mapping Requirements

The selected requirements statements from the previous step are mapped into ontology concept. As mentioned in the previous section, the ontology concept used in this research is a combination of the general requirements ontology and domain application ontology. In order to link the related concepts from both ontologies, several new properties are created. Table 1 shows the properties that are used for linking those concepts. This work keeps the properties and classes that exist in both of general requirements ontology and domain application ontology.

Table 1. Property for lingking ontology concept

Property name	Domain	Range
Relate_AandS	Actor	ont:Stakeholder
Relate_ATandS	ActorType	ont:Stakeholder
Relate_AandR	Application	ont:Requirement
Relate_DandR	Domain	ont:Requirement
Relate_DandRA	Data	ont:Requirement-Artifact

Combining two different ontologies helps to interpret the requirements of a complex system in a more comprehensive point of view. The requirements ontology helps to describe the desired software characteristic. Moreover, domain ontology represents the application specific domain knowledge. This combined ontology also provides efficient searching and browsing materials that are able to detect changes and maintain requirements knowledge. By using the combined ontology, the process to trace back requirements based on its goal determined by stakeholder and the environment becomes easier. In the end of the mapping requirements process, requirements analyst is asked to validate the mapping result.

4 Case Study and Implementation

4.1 Smart Grid System

U.S. Department of Energy in [23] defines smart grid system as a fully automated power delivers a network that monitors and controls every customer and node, ensuring a two-way flow of the electricity and the information between the power plant and the appliance, and the entire points. A complex system like smart grid system is related to various domains like as a market, a customer, a provider and an operation that has different user viewpoints. Even though each domain has an independent environment, their roles or policies are strongly related. In this situation, stakeholders should consider dependency between operators or functions in the development process. Therefore, when the requirements of the system are changed, the related functionality should be examined carefully. Due to the smart grid system grows with the size and the complexity; it is not easy to recognize the association between requirements. In order to solve that matter, managing requirements become an important issue for smart grid.

In this research, we aim to provide management in modeling requirements in a specific system. In order to get the requirements asset, we use scenarios that are gathered from [24, 25], for smart grid system. As our case study we gathered two types of scenario based on its types of smart grid which are micro smart grid and smart grid. The first scenario is about home grid that focuses on the heating system. The smart home system should react based on the temperature of the environment. The heating devices should be turned off if the temperature reaches into the minimum constrains. The second scenario shows that each home tries to minimize its electricity bill, but still gets the maximum energy. Therefore, we can see that the risen issue is related to balancing the interest of the different parties involved.

4.2 Linking Requirements Ontology and Smart Grid Ontology

In order to ensure traceability using ontology, we start our work by determining the ontology for requirement. This ontology is used to map the concept that will be used to identify the specification of the system. In this process, we use ontology for requirement built by Siegemund in [26] seen in Fig. 3.

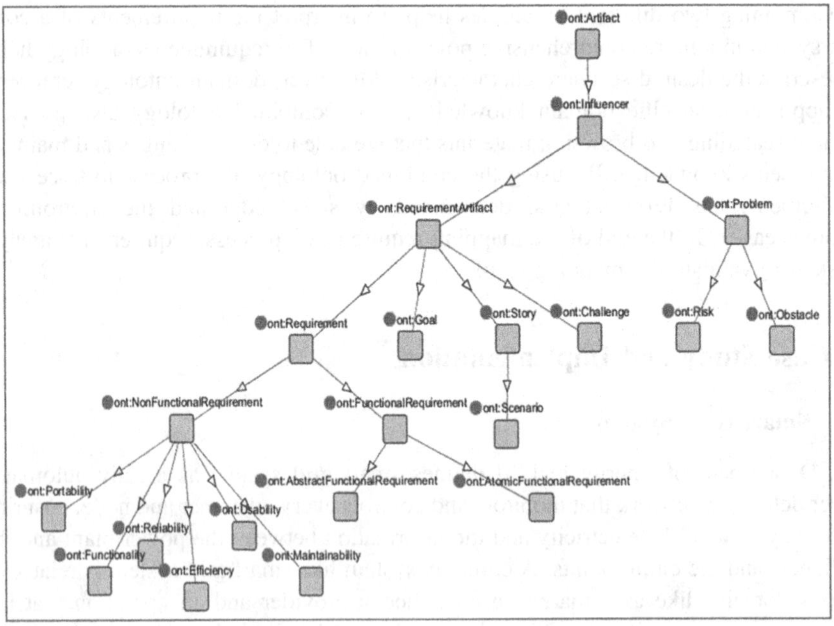

Fig. 3. Hierarchical view of requirements ontology

The class *RequirementArtifact* is the most important concept in this ontology. An action or a behavior of system is represented in sub-concepts of the *Requirement*. Then, the limitation of the system is represented in *Problem*. Moreover, their objective can be included in the *Goal*. The relations among the concept are also determined during this process. These relations help to identify as the association between the functions of the system. The other main concept is *Stakeholder* which does not have sub-concept. The *Stakeholder* includes users, developer, manager. This concept is one of the important because for managing the evolving requirements, we should able to determine the origin of the requirements. *Functional* and *non-functional requirement* of the system will be determined under the *Requirement* class.

The *functional* requirement will be described as abstract functional and atomic functional requirement. This ontology also covers about *non-functional requirement* includes *efficiency, functionality, maintainability, portability, reliability,* and *usability.* Efficiency class sub classes which is compliance, resource, and time behavior. For the functionality the subclass includes *accuracy, interoperability, security,* and *suitability.* Then for the *maintainability* includes *analyzability changeability, stability* and *testability.* The fourth subclass in *non-functionality* class is *portability* that includes *adaptability, coexistence, conformance, install ability,* and *replace ability.* Then for the reliability includes fault tolerance, maturity, and recoverability. The last one is *usability* that includes *attractiveness, effectiveness, learnability, operability, understand ability* and *user satisfaction.*

By collecting data from several resources we built the ontology for the smart grid. There are several classes in our smart grid ontology. The First one is *actor* which has the capability to make decisions and exchange information with other actor. The *actor* can be devices, computer system or the organization. Each of the actors will take a certain role in the smart grid system that will be represented in the action *role* class.

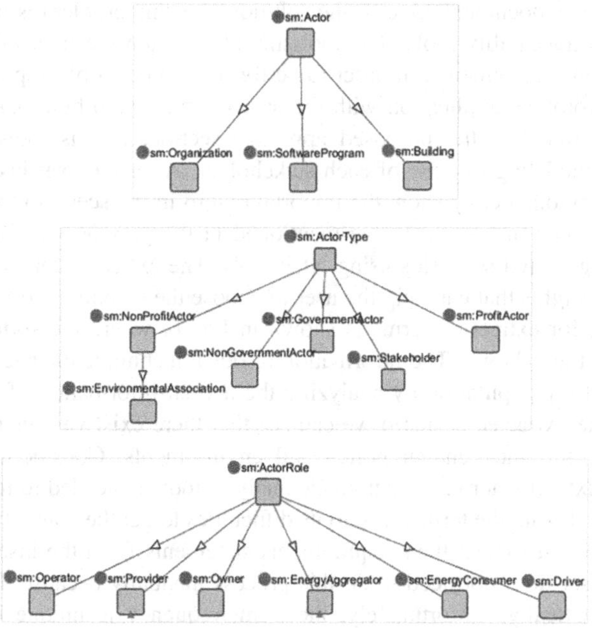

Fig. 4. The snippet of main concept and sub-concept of the extended ontology derived by combining requirements and smart grid ontology

Domain class is also important in the smart grid concept. Understanding the domain that correlated in this system is important because the smart grid system is a complex system that related with more than one area.

The main concepts in the combined ontology are *Actor*, *Application*, *ActorRole*, *Domain* and *Type*. Figure 4 shows some portion of the main concepts and their sub-concepts. The *Actor* has the capability to make decisions and exchange information with other actors through interface. Each of the *Actor* will take a certain role in the smart grid system that will be represented in the *ActorRole*. We only consider the *ActorType* in Type concept. The Application concept represents the tasks which are performed by *Actor*. We linked the requirements ontology and the smart gird ontology by using the defined properties. The *ActorType* describes types of actors such as stakeholder in smart grid system. The *ActorType* is linked by the *relatedToStakeholder* property. Moreover, *ActorRole* is related to Goal therefore this concept is linked by the *relatedToGoal*. Application can be linked to *Requirement* or *Goal*. Then, *Problem* also is linked to *Application* because *Application* is associated with a certain condition or capability.

4.3 Implementation and Discussion

Requirements traceability helps to identify whether the entire requirements have been implemented consistently. However, the task for providing manual requirements traceability tends to become costly, time consuming, and error-prone procedure. It is unwise to invest the resources for manual monitoring and updating the traceability link in the requirements document. One of the solutions for this problem is the production of requirements traceability tools. The tool should maintain the traceability based on the links between requirements. In order to perform the process of proposed approach, we provide a prototype application with the user interface (UI) built using Java.

As we mentioned in the proposed approach section, in this tools, we use the scenario based modeling to need of each stakeholder. After the user inserted the scenario, the tools automatically show the important word in the scenario by highlighting the keyword as seen in Fig. 5. As we mentioned in the previous section, we extract terms from the gathered scenarios using NLP tools. The extracted terms are presented based on their weights that can help the user to choose the essential terms The result of the text parsing for extracting terms is shown in Fig. 6, where the sorted terms, tag, NER and weight are shown. The information retrieval technique enhances the process during the ontology population by analyzing the textual information of the artifacts.

Based on the extracted scenario we can see that there exist various users in smart grid. Each user has independent policy and environments. Considering of the big number of the extracted term, the automatic analysis tools is needed to reduce the cost and effort for analyzing the terms. Due to its difficulties to get the context of each word, the users are asked to select their requirements statements from the inserted scenario. However, the user does not need to do this process manually because the keyword is extracted automatically. Unfortunately, the term frequency technique is not able to analyze the term based on their semantic meaning. The extraction process in our proposed approach is done by considering the frequency of the term. In order to

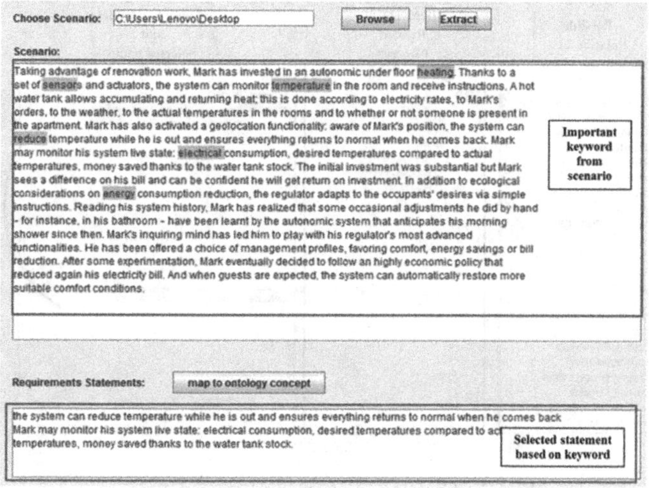

Fig. 5. Defining requirements statements

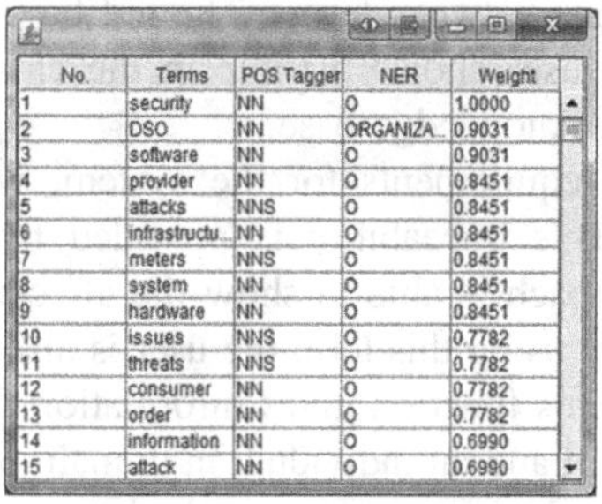

No.	Terms	POS Tagger	NER	Weight
1	security	NN	O	1.0000
2	DSO	NN	ORGANIZA...	0.9031
3	software	NN	O	0.9031
4	provider	NN	O	0.8451
5	attacks	NNS	O	0.8451
6	infrastructu...	NN	O	0.8451
7	meters	NNS	O	0.8451
8	system	NN	O	0.8451
9	hardware	NN	O	0.8451
10	issues	NNS	O	0.7782
11	threats	NNS	O	0.7782
12	consumer	NN	O	0.7782
13	order	NN	O	0.7782
14	information	NN	O	0.6990
15	attack	NN	O	0.6990

Fig. 6. The extracted terms and its weight based on term frequency

overcome this limitation, we asked the user to select their requirements statements based on the extracted term by clicking the word.

In the next process the user is also asked to validate the result of mapping the term into ontology concept. The mapping process in our proposed approach is done by considering the properties of the extracted term including domain, relation, range. The central concept of our tools is the stakeholder and abstract requirement as well as the property details and defines. The abstract will have some subclasses. They are goal, scenario and requirement. Each of stakeholders will define all information in details by other abstract requirements. Combining two different ontologies helps us to interpret the requirements of a complex system in a more comprehensive point of view.

After each term is assigned into a certain concept, the next step is defining properties. When a term is dragged into a concept, at the same time, the new window is opened with the possible relation list as seen in Fig. 7. This window is used to define a relation with another asset in the domain ontology. When the possible relation list in Fig. 7 is selected, the user can see related concepts. A user can make relation between

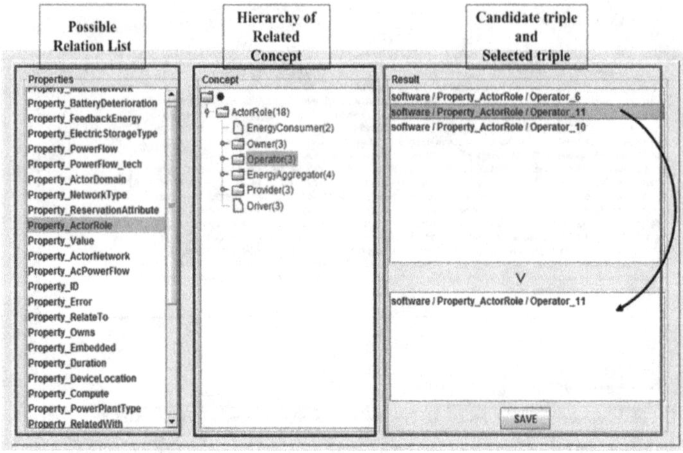

Fig. 7. Creating relationship between term

terms by selecting a candidate triple (domain, relation, range). The prototype makes the user is able to create relations between terms. The red box shows the possible list to create a relation and when user choice a triple in the candidate triple list, selected candidate triple is created in knowledge base.

After we got the user requirements for the system, we work for handling the requirements traceability. This traceability is provided by the link from ontology concept. Our proposed approach is able to show the affected classes when there is a change in a certain class. By using this tool, the user is also able to search the related artifacts as seen in Fig. 8. This function gives information through the relation. If the user selects a specific concept and an individual in domain ontology, the tool provides related concepts and individuals that are associated with selected concept or individual. This function makes it possible that when any artifact is modified, the user can identify the influenced artifacts.

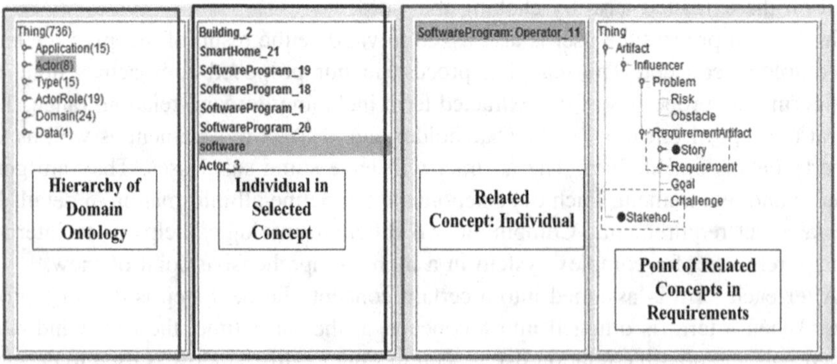

Fig. 8. Searching related concepts

By showing the affected class, the user is able to consider the other class that should be changed due to their relations with the changed class. In our current work, the change that can be traced is the previously generated requirements. So, the user only can change an existing mapped requirement. If there is a case that the user wants to change the scenario, the user should do the remapping process. Using this function, it will be easier for the user to identify the effect of changing, requiring and tracing the relationship between requirements.

5 Conclusion and Future Work

One common problem usually faced during system development is the changing environments which results in the possibility of evolving requirements. This problem can be solved using the requirements traceability. Requirements traceability helps to identify whether the entire requirements have been implemented consistently. However, manually providing requirements traceability could be time-consuming and error-prone.

Our proposed approach is able to solve this problem by managing the evolving requirements. Furthermore, the result of our proposed approach shows the relationship of the requirements generated using ontology-based approach. The key contribution of our proposed approach is the ability to reduce the effort to identify the evolving requirements. Using our proposed approach, the requirements analyst can determine the requirements of users that are described through scenarios. These requirements are mapped into the combination of requirements ontology and domain to get the relation of the requirements artifact. In conclusion, this mapping process helps users to trace the affected requirements if there is a change in certain requirements.

Due to the limitation of the information retrieval technique used in this paper, users need to analyze the extracted term manually. Further work will focus on the implementation of learning process to provide a fully automatic mapping process in order to reduce time and human resources. Moreover, relying on term frequency technique does not guarantee the quality of the traceability, the further research needs another method that is able to fill the "semantic" gap. Therefore, our future work also includes the knowledge acquisition technique, which is able to analyze the term semantically.

Acknowledgement. This research was supported by Next-Generation Information Computing Development Program through the National Research Foundation of Korea (NRF) funded by the Ministry of Science, ICT & Future Planning (2013M3C4A7056233).

References

1. Ali, N., Gueneuc, Y.G., Antoniol, G.: Trustrace: mining software repositories to improve the accuracy of requirement traceability links. IEEE Trans. Softw. Eng. **39**(5), 725–741 (2013)
2. Nuseibeh, B., Easterbrook, S.: Requirements engineering: a roadmap. In: Proceedings of the Conference on the Future of Software Engineering, pp. 35–46. ACM (2000)

3. Gotel, O.C., Finkelstein, A.C.: An analysis of the requirements traceability problem. In: Requirements Engineering Conference, pp. 94–101. IEEE (1994)
4. Cuddeback, D., Dekhtyar, A., Hayes, J.H.: Automated requirements traceability: the study of human analysts. In: 2010 18th IEEE International Requirements Engineering Conference (RE), pp. 231–240. IEEE (2010)
5. Richter, H., Gandhi, R., Liu, L., Lee, S.W.: Incorporating multimedia source materials into a traceability framework. In: Proceedings of the First International Workshop on Multimedia Requirements Engineering, p. 7. IEEE, September 2006
6. Heindl, M., Biffl, S.: A case study on value-based requirements tracing. In: Proceedings of the 13th ACM SIGSOFT International Symposium on Foundations of Software Engineering, pp. 60–69. ACM, September 2005
7. Kannenberg, A., Saiedian, H.: Why software requirements traceability remains a challenge. CrossTalk J. Defense Softw. Eng. **22**(5), 14–19 (2009)
8. Davis, A.M.: Software Requirements: Analysis and Specification. Prentice Hall Press, Upper Saddle River (1990)
9. Egyed, A., Grunbacher, P.: Automating requirements traceability: beyond the record & replay paradigm. In: Proceedings of the 17th IEEE International Conference on Automated Software Engineering, ASE 2002, pp. 163–171. IEEE (2002)
10. Naslavsky, L., Alspaugh, T.A., Richardson, D.J., Ziv, H.: Using scenarios to support traceability. In: Proceedings of the 3rd International Workshop on Traceability in Emerging Forms of Software Engineering, pp. 25–30. ACM, November 2005
11. Riechert, T., Lauenroth, K., Lehmann, J., Auer, S.: Towards semantic based requirements engineering. In: Proceedings of the 7th International Conference on Knowledge Management (I-KNOW) (2007)
12. Chowdhury, G.: Introduction to Modern Information Retrieval. Facet Publishing (2010)
13. Luhn, H.P.: A statistical approach to mechanized encoding and searching of literary information. IBM J. Res. Dev. **1**(4), 309–317 (1957)
14. Hayes, J.H., Dekhtyar, A., Osborne, J.: Improving requirements tracing via information retrieval. In: Proceedings of the 11th IEEE International Requirements Engineering Conference, pp. 138–147. IEEE (2003)
15. Cleland-Huang, J., Guo, J.: Towards more intelligent trace retrieval algorithms. In: Proceedings of the 3rd International Workshop on Realizing Artificial Intelligence Synergies in Software Engineering, pp. 1–6. ACM (2014)
16. Noy, N.F., McGuinness, D.L.: Ontology development 101: A guide to creating your first ontology (2001)
17. Menzies, T.: Cost benefits of ontologies. Intelligence **10**(3), 26–32 (1999)
18. Assawamekin, N., Sunetnanta, T., Pluempitiwiriyawej, C.: Ontology-based multiperspective requirements traceability framework. Knowl. Inf. Syst. **25**(3), 493–522 (2010)
19. Zhang, Y., Witte, R., Rilling, J., Haarslev, V.: An ontology-based approach for traceability recovery. In: 3rd International Workshop on Metamodels, Schemas, Grammars, and Ontologies for Reverse Engineering (ATEM 2006), Genoa, pp. 36–43 (2006)
20. Narayan, N., Bruegge, B., Delater, A., Paech, B.: Enhanced traceability in model-based CASE tools using ontologies and information retrieval. In: 2011 Fourth International Workshop on Managing Requirements Knowledge (MARK), pp. 24–28). IEEE (2011)
21. Manning, C., Jurafsky, D., Liang, P.: The Stanford Natural Language Processing Group (2010)
22. Toutanova, K., Klein, D., Manning, C.D., Singer, Y.: Feature-rich part-of-speech tagging with a cyclic dependency network. In: Proceedings of the 2003 Conference of the North American Chapter of the Association for Computational Linguistics on Human Language Technology, vol. 1, pp. 173–180. Association for Computational Linguistics (2003)

23. Doe, U.: Grid 2030: A national vision for electricity's second 100 years.US DOE Report (2003)
24. Frey, S., Huguet, F., Mivielle, C., Menga, D., Diaconescu, A., Demeure, I.M.: Scenarios for an autonomic micro smart grid. In: SMARTGREENS, pp. 137–140 (2012)
25. Barenghi, A., Bertoni, G.M., Breveglieri, L., Fugini, M.G., Pelosi, G.: Smart metering in power grids: application scenarios and security. In: 1st IEEE PES Innovative Smart Grid Technologies (ISGT) Asia Conference, Perth (2011)
26. Siegemund, K., Thomas, E.J., Zhao, Y., Pan, J., Assmann, U.: Towards ontology-driven requirements engineering. In: Workshop Semantic Web Enabled Software Engineering at 10th International Semantic Web Conference (ISWC), Bonn (2011)
27. Aponte, J., Marcus, A.: Improving traceability link recovery methods through software artifact summarization. In: Proceedings of the 6th International Workshop on Traceability in Emerging Forms of Software Engineering, pp. 46–49. ACM (2011)

Situation-Oriented Evaluation and Prioritization of Requirements

Nimanthi L. Atukorala[1(✉)], Carl K. Chang[1], and Katsunori Oyama[2]

[1] Department of Computer Science, Iowa State University, Ames, USA
{nimanthi, chang}@iastate.edu

[2] Department of Computer Science, Nihon University, Koriyama, Japan
oyama@cs.ce.nihon-u.ac.jp

Abstract. Evaluation and prioritization of requirements is one of the key aspects in requirements engineering. Although the existing studies in this area are greatly focused on addressing business goals such as development budget and deadlines of completion, we believe that human-centered concerns, including end-users' personal desires, goals, beliefs and constrained environment, must also weigh in. In this paper we present a new human-centered requirements evaluation and prioritization approach that effectively considers such concerns. The proposed method is based on the situation–transition structure introduced in our previous study that was used to elicit human-centered requirements. We illustrate the applicability of the proposed methodology through a case study.

Keywords: Evaluation · Prioritization · Requirements · Human-centered · Situation · Situation–transition structure

1 Introduction

High end-user satisfaction is always the ultimate goal of any software development project. Fulfilling the right set of end-users' requirements is the key to achieving this goal. However, finding that right set of end-users' requirements is challenging. Various requirements prioritization techniques have been introduced to support this task mainly based on aspects such as their importance for the overall system functionality, limits on budget, required time, potential risk, etc. [1].

Each individual human being is unique and has his or her own unique interests, desires, personal goals, etc. Moreover, their origin, culture and surrounding environmental factors lead them to have their own unique beliefs. Individuals' requirements are molded by these unique human factors and surrounding environmental factors. When it comes to software development, each individual end-user has a set of prioritized requirements imposed upon the particular software, and the satisfaction depends on whether or not those requirements are fulfilled by the software. It is also important to note that the requirements that have higher priority to one end-user can be entirely unimportant to another. Based on these observations, we believe that the requirements prioritization process can be made more effective by considering the information about the uniqueness of an individual's prioritization of those requirements [2]. In this paper we

S.-W. Lee and T. Nakatani (Eds.): APRES 2016, CCIS 671, pp. 18–33, 2016.
DOI: 10.1007/978-981-10-3256-1_2

propose a new approach to prioritizing requirements on an individual basis by considering user's human nature. The requirements are evaluated based on two main factors:

1. The importance of the requirement to a particular end-user (encoded as the Importance factor)
2. The likelihood that the requirement leads to an undesirable or risky situation (encoded as the Risk factor)

According to the proposed requirements prioritization approach a requirement will be assigned higher priority if its Importance factor is high and the Risk factor is low. The proposed requirements prioritization approach is an extension to our previous study on eliciting new human-centered requirements [3]. In a previous study we introduced a situation-transition structure that represent the human behavioral patterns in a computational format, and used it for requirements elicitation. The proposed requirements prioritization technique uses the information included in that situation-transition structure. In addition, we also extend the computerized system developed in the previous study so that it can be used to identify and prioritize requirements for each individual. It is important to note that the proposed requirements prioritization approach is not aiming to completely eliminate the use of other approaches, but instead provide additional information to the requirements analysts to make better decisions during the requirements prioritizing phase.

The rest of this paper is organized as follows: Sect. 2 reviews previous studies about requirements evaluation and prioritization. Section 3 summarizes the concept of situations, situation-transition structure and reviews the requirements elicitation procedure proposed in our previous study. Section 4 describes the proposed method in detail. In Sect. 5 a summary of a real-life case study is given for illustration. Section 6 includes a discussion and finally, Sect. 7 concludes the paper with some future work suggestions.

2 Related Work

The fast growing software development industry strives to release new products and their enhancements as frequently as possible. In order to reduce the problems starting from resource limitations and time constrains and maximize revenue, many software development companies pay more attention to the evaluation and prioritization of requirements than ever before. The rest of this section will discuss some of the common techniques as well as some studies in requirements evaluation and prioritization.

2.1 Direct Stakeholder Collaboration Based Approaches

Most of the traditional requirements prioritization techniques allow the stakeholders to prioritize the requirements according to their personal preferences and then form an agreement through identifying conflicts.

Numerical assignment is the most common approach where stakeholders are requested to place requirements in priority groups [4]. The number of priority groups depends on the software development practice, but three groups: critical, standard, and optional are very common [5]. Win-win, also known as Theory-W [6], is a prioritization technique that allows each stakeholder to categorize requirements according to importance and potential risk, whereas the Top-Ten requirements approach [1] allows stakeholders to pick their top-ten requirements from a larger set without assigning an internal order between the requirements. In a 100-Dollar Test [1], each stakeholder is given 100 imaginary units (such as money, hours, etc.) to distribute among requirements and consider the ratio of the assignment as the scale of the priority. Although most of these techniques are simple and easily manipulated by stakeholders, each has its own limitations. One common problem in these approaches is that most of the stakeholders think that everything is rather critical. For example, a study [7] shows that stakeholders most likely consider 85 percent of the requirements as critical.

The proposed approach does not take direct responses from the end-users, but instead uses the observational data to make an unbiased decision on requirements prioritization. We believe that this will be an effective alternative method to discern the actual critical requirements of the end-users. In addition, each requirement will be assigned a unique priority, which will also be helpful for better decision making. Our proposed approach can be considered as a special case of win-win prioritization technique, where the requirements are prioritized based on their relative importance and the potential risks to the stakeholders.

2.2 Search Based Approaches

A well-known search based requirements prioritization approach applies Binary Search Trees (BST) where the prioritization is performed by constructing a binary search tree so that less important requirements are inserted to the left and more important ones to the right. A list of ranked requirements is obtained by using the bubble sort or binary search tree algorithms [1]. This allows stakeholders to compare the relative value of individual requirements and can be used to prioritize relatively large sets of requirements [6]. However, it is believed that original BST ranking is more suitable for a single stakeholder regardless of the sorting algorithm used for ranking since aligning several different stakeholders' views at the same time might be difficult [1].

Bebensee et al. [8] introduced Binary Priority List (BPT), a variation of BST structure, for prioritizing software requirements. The level of a requirement in the BPT represents its priority level. The top-most level has the highest priority and the priority decreases from top to bottom. Beg et al. [9] proposed requirements prioritization technique using B-tree, a self-balancing tree data structure aiming to reduce the number of comparisons between requirements pairs.

In comparison with BST, the situation-transition structure used in the proposed approach is a complex graph with set of nodes, directed edges and weight values on edges. The unique rank for each requirement is obtained using a graph traversal algorithm developed by modifying the depth first search algorithms. In other words, the procedure of the proposed approach is similar to BST ranking; however, the proposed approach can be used in both single and multiple end-user domains. Therefore, it is more powerful.

2.3 Machine Learning Based Approaches

Some researchers are focused on applying data mining and machine learning techniques to the requirements prioritization process in order to improve their effectiveness in large software development projects with multiple stakeholders. In [10], Duan et al. proposed a Pirogov approach that uses clustering techniques to place requirements into multiple independent clusters that capture the diverse and complex roles played by individual requirements. Stakeholders determine the relative value of each cluster and weight the importance of each clustering method. An objective function then generates prioritization decisions at the level of the individual requirement.

Tonella et al. [11] proposed an Interactive Genetic Algorithm (IGA) that includes incremental knowledge acquisition and combines it with the existing constraints, such as dependencies and priorities. This approach aims at minimizing the disagreement between a total order of prioritized requirements and the various constraints that are either encoded with the requirements or that are expressed iteratively by the user during the prioritization process. An interactive genetic algorithm was used to achieve such a minimization, taking advantage of interactive input from the user whenever the fitness function cannot be computed precisely based on the information available. The process terminates when a low disagreement is reached, the time out is reached or the allocated elicitation budget has been consumed.

Perini et al. [12] proposed a requirement prioritization method based on Case-Based Ranking (CBRank). CBRank originated from a framework that supports decision making on ordering a set of items which can handle single and multiple stakeholders and different ordering criteria. The prioritization is performed by considering the stakeholders' preferences and the approximated ordering of requirements predicted by machine learning techniques. Similarly, Babar et al. [13] proposed the PHandler, which is an intelligent requirements prioritizing technique that uses artificial neural networks to predict the priority of the requirements.

The algorithm used to generate the situation-transition structure in our proposed approach is a modified version of Chow-Liu Bayesian structure learning algorithm [14], which is popular in the machine learning area. In other words, the proposed approach also uses machine learning techniques to predict the priority of the requirements based on the end-user's observational data.

2.4 External Factors

Some recent research on stakeholder based requirements prioritizations are focused on enhancing the reliability of the approaches. For example, Ahmad et al. [15] discusses limitations of existing requirements prioritization techniques with respect to geographical distribution of stakeholders and provides a framework to identify important requirements of a product to be useful for distributed development. As this recent study suggested, we also believe that consideration of end-users' human nature and environmental factors such as geographical distribution will improve the quality of the requirement prioritization. Note that the basic computational unit *situation* used in our proposed approach is representing information about both aspects.

3 Situations, Situation-Transition Structure and Requirements Elicitation Using Situation-Transition Structure

3.1 Situations

Situ defined in [16] is a computational framework that describes the process of inferring human desires through relevant human actions and the environmental contexts. The term "*situation*" defined in the Situ framework encapsulates these three factors: human desire, actions and environmental context are a single computational unit. Our proposed methodology uses this definition of situation in order to embed human factors as well as origin, culture and surrounding environmental factors into the situation-transition structure.

Definition: A situation at time t, is a 3-tuple {d, A, E}$_t$ in which d is the predicted end-user's desire, A is a set of end-user's actions to achieve a goal which d corresponds to, and E is a set of environmental context values with respect to a subset of the context variables at time t.

According to the above definition of situation, it is possible to pair time stamped records of desire, set of actions and environmental context values of a particular person into situation tuples which leads to a *sequence of situations* over time. Moreover, it is possible that a person may have multiple desires at a given time *t* and actions to be performed during that time to reflect one or more of those multiple desires. In other words, a sequence of situations of an individual may include time periods where multiple situations occur simultaneously (a set of concurrent situations).

3.2 Situation-Transition Structure

One significant observed property in the sequence of situations of an individual is that some situations are more likely to transition to a specific subset of future situations than others. We call this property *situation transition.* Hence the term *situation-transition structure* in this paper refers to a domain specific directed weighted graph generated using all possible situation transitions of a person in a particular domain during a pre-defined time period. The overall process of deriving situation-transition structure from observational data is given in [3].

Figure 1 shows an example of such situation transition structure. Here, each node represents a possible situation or a set of concurrent situations in the sequence. An edge from node X to Y represents that in the observational data situation, or a set of concurrent situations represented by node X, is more likely to transit to the situation or a set of concurrent situations represented by node Y. Each edge is annotated with a mutual information value. The mutual information of the directed edge from node X to node Y, I(X,Y) is calculated using formula (1). Suppose X,Y represent situations (or a set of concurrent situations) SituX, SituY respectively.

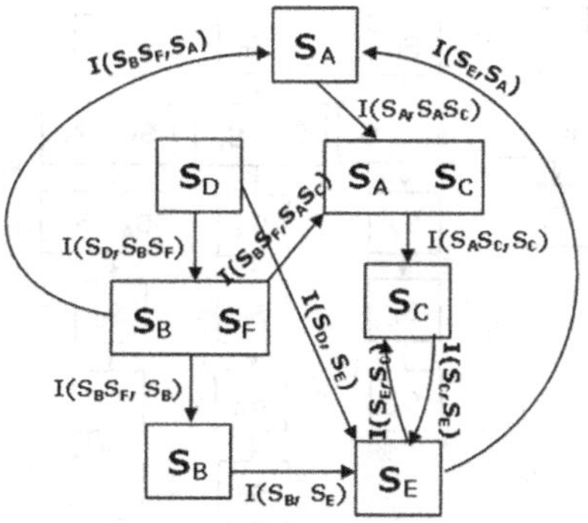

Fig. 1. An example situation-transition structure

$$I(X,Y) = \sum_{x \in values\,(X)} \sum_{y \in values\,(Y)} P(x,y) \log_2 \left(\frac{P(x,y)}{P(x)P(y)} \right) \quad (1)$$

where,
values(X) = {True, False}
True: When X is the parent node of the pair. i.e. SituX occurred just before SituY
False: When X is not the parent node of the pair. i.e. SituX does not occur just before SituY
values(Y) = {True, False}
True: When Y is the child node of the pair. i.e. SituY occurred just after SituX
False: When Y is not the child node of the pair. i.e. SituY does not occur just after SituX
P(x,y) = Joint probability of X = x and Y = y
P(x) = Probability of X = x (0 < P(x) < 1)
P(y) = Probability of Y = y (0 < P(y) < 1)

Note that the mutual information is directly proportional to the likelihood of this transition. That is, if the mutual information is high, the likelihood of the transition is also high and vice-versa.

3.3 Requirements Elicitation Using Situation-Transition Structure

The procedure of eliciting new requirements from situation-transition structure starts with rearranging the structure into a rooted structure by selecting the appropriate node as the root. Selection of root depends on the domain and set of available situations. We assume that the requirements analyst manually performs this task. Usually, the node representing the initial situation of the sequence of situations or the situation where all the actions and environmental context are in their default values is selected as the root. In cases where all the possible paths in the situation-transition structure cannot be

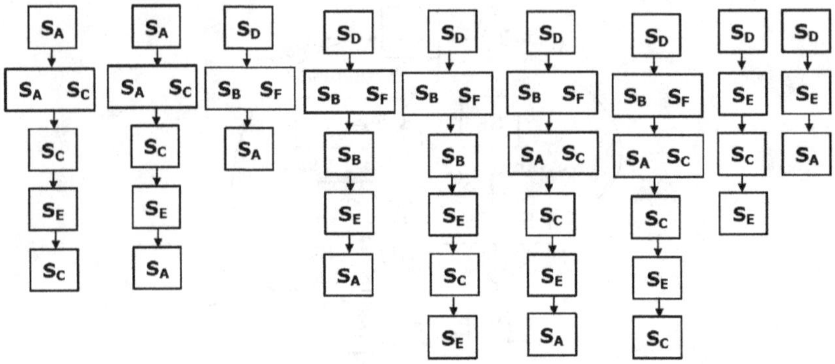

Fig. 2. Set of possible paths in situation-transition structure in Fig. 1.

traversed by starting from a single root node, it is required to select multiple root nodes. Once the root nodes are selected, all the possible paths starting from the root are listed. Figure 2 shows the set of possible paths in the situation-transition structure given in Fig. 1. Here S_A and S_D are selected as the root nodes.

In our earlier study [3] on feature extraction terms and the new requirements elicitation procedure, we proposed five feature extraction terms: direct cause, leads to, terminate, sustain, and prevent in order to analyze the situation-transition structure paths. These five feature extraction terms are then used to define some requirements construction terms so that the elicited new requirements can be presented in semi-formal manner.

4 Requirements Evaluation and Prioritization Using the Situation-Transition Structure

As mentioned earlier, the proposed requirements prioritization approach is based on two main factors: Importance factor and Risk factor. The rest of this section explains each of these two factors.

4.1 Importance Factor of Requirements

According to the definition, the Importance factor in the proposed approach measures the importance of the requirement to the end-user. Here we assume that the importance or relevance of the requirement to the end-user is directly proportional to how often the user requires the feature provided by that requirement. As given in Sect. 3, the elicited requirements are related to one or more situation transitions in the situation-transition structure. Therefore, based on our previous assumption, it is possible to claim that the importance of a particular requirement is directly proportional to the likelihood that the user is engaged in the situation transitions which leads to eliciting that particular requirement. As described in Sect. 3.2, the mutual information can be used as a quantity that represents the likelihood of this transition. Hence, we can use mutual information of the transitions to evaluate the Importance factor of a particular requirement.

1. *Group Importance Factor*

As defined earlier, each path in the situation-transition structure is a collection of edges where each edge annotated with a mutual information value that represents the likelihood of the transition between the two nodes. We define the path frequency indicator factor as follows:

Definition: The path frequency indicator factor of a path p (pathFreq(p)) is the sum of mutual information values of the associated situation transitions.

$$pathFreq(p) = \sum_{\substack{<X,\, Y> \; \in \; Set\, of\, transitions \\ associated\, with\, path\, p}} I(X, Y) \qquad (2)$$

where, I(X,Y) = Mutual information value of the transition from X to Y

First, we calculate the path frequency indicator factors of the possible paths identified in the situation-transition structure. Next, we arrange the paths in the situation-transition structure according to the descending order of the path frequency indicator factor. Then, we find the group of requirements for each ordered path such that a requirement belongs to at most one group. For each group, assign an Importance factor within a pre-defined range decided by the requirements analyst such that groups where a path related to higher path frequency indicator factor gets a higher Importance factor and the groups with a lower path frequency indicator factor gets a lower Importance factor. Note that the requirements in a particular group are assigned the same Importance factor.

2. *Individual Requirement Importance Factor*

The situation-transition structure can also be used to get the Importance factor for each individual requirement as described follows: Consider a requirement *R*. Let *p* denote the path in the situation-transition structure that leads to eliciting *R* and *I(X,Y)* denote the mutual information value of a transition <X,Y> in the path. Then, Importance factor of *R*, IMP_R can be defined as:

$$IMP_R = \sum_{\substack{<X,\, Y> \; \in \; Set\, of\, transitions \\ associated\, with\, path\, p\, that\, is\, related\, to\, R}} I(X, Y) \qquad (3)$$

4.2 Risk Factor of Requirements

In general the term risk in requirements prioritization refers to the risks involved in business and common issues such as technical or market risks, violation of regulations, unavailability of suppliers, etc. [1]. However, in this paper the term *risk* is used to represent the possible risks related to end-users. We believe that the derived situation-transition structure can be used to gain knowledge about such potential risks.

Note that sometimes end-users' behaviors can result in situations with undesired or harmful outcomes. We call such situations *risk situations* or *hazard situations*. Conversely, some other end-users' behaviors can result in un-harmful situations. We call such situations as *safe situations*. Moreover, some safe situations can also mitigate the possible damage from risk situations. We call those situations *safe defender situations*. Note that set of safe defender situations is contained inside the set of safe situations.

In an ideal case, all the situations in the situation-transition structure must be labeled as either safe or risk situation. However, in practice there can be situations that cannot be clearly labeled as safe or risky. Such situations are called as *un-labeled situations*.

1. *Identifying Three Categories of Situations*

The first step is to label each node in the situation-transition structure using one of the three categories: risk, safe, safe defender. The label is also supplemented with scalar that represents the relative risk, safe and defender. The scalar is defined within the range +5 to −5 as follows:

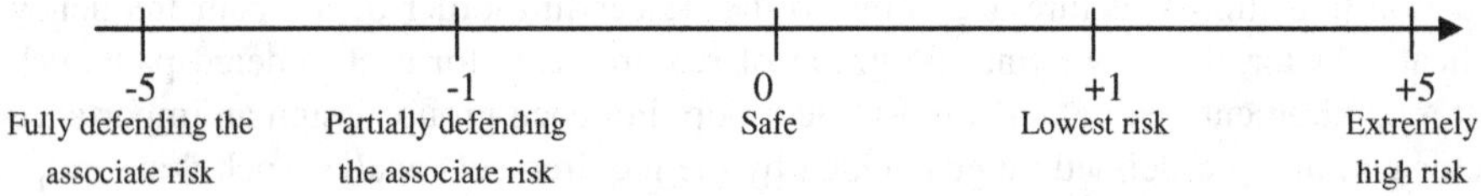

We assume that a requirement analyst with some background knowledge can perform this labeling task. If the requirement analyst is unable to label a situation, it will be remain as un-labeled situation.

The next step is to identify the mapping between risk situations and the corresponding safe defender situations. Note that, this is not a "one-to-one" mapping. It is possible that there exists a risk situation that can be mapped to more than one safe defender situation while another risk situation may not be mapped to any such safe defender situation at all. However, we assume that the mapping is "on-to" which implies that each safe defender situation is mapped to at least one risk situation. If a safe defender situation is mapped to more than one risk situation, the requirement analyst can label it with multiple scalar values for each risk. Figure 3 visualizes the three categories and the mapping.

2. *Finding Risk Factor*

Consider a requirement *R*. Let *S* denote the situations related to the transitions that leads to eliciting R. Then, the Risk factor of *R*, $RISK_R$ can be defined as,

$$RISK_R = \sum_{i \in S} Assigned\ Scalar\ of\ i \tag{4}$$

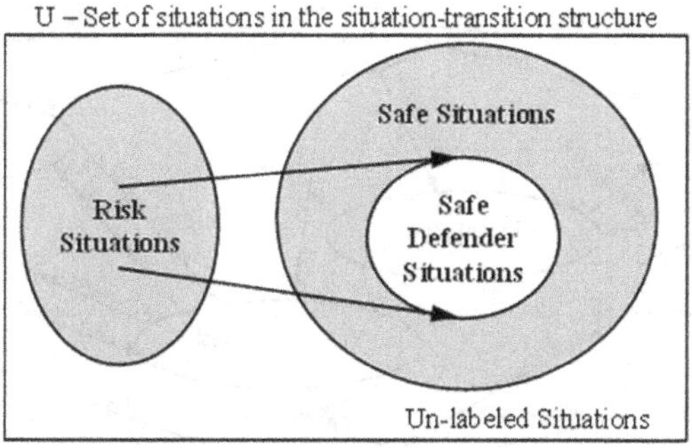

Fig. 3. Three categories of situations: safe, risk and safe defender and the mapping

4.3 Requirements Prioritization

According to the proposed approach the priority of a requirement *R* is increasing when its Importance factor (IMP_R) is increasing or the Risk factor ($RISK_R$) is decreasing. Hence, the priority of the requirement *R* (*Priority(R)*) can be estimated as follows:

$$Priority(R) = \begin{cases} k\dfrac{IMP_R}{RISK_R}\ \mathit{If}\ RISK_R > 0 \\ kIMP_R\ \mathit{Otherwise} \end{cases} \quad (5)$$

where, k is a positive constant.

A requirements analyst can select positive constant k such that the priorities values lies within a pre-defined range. Note that it is possible to perform both requirements elicitation and prioritization in parallel, so that we can directly elicit a prioritized set of requirements. This will greatly reduce the time and workload of the process.

5 Cooperative Research Environment (CoRE) – A Case Study

In this section we present early stage results of a case study we have conducted to show the applicability of the proposed requirements prioritization approach. CoRE is a website for sharing published and unpublished internal research papers, comments and ideas in a research environment. The existing version of the CoRE website was developed based on a set of requirements elicited by the requirements analysts through traditional brainstorming techniques with no involvement of end-users within the process. Our goal is to find a list of prioritized requirements for the next version of this website through the end-users' observational data.

We considered the current version of the website as the domain and the participants of the website as the end-users. We have collected the records on 65 end-user actions on the current CoRE website interface, such as button and link clicks, menu option selections, current and next webpages (10062 records within 34 days) [17]. In this dataset the actions of each end-user can be uniquely identified through their login

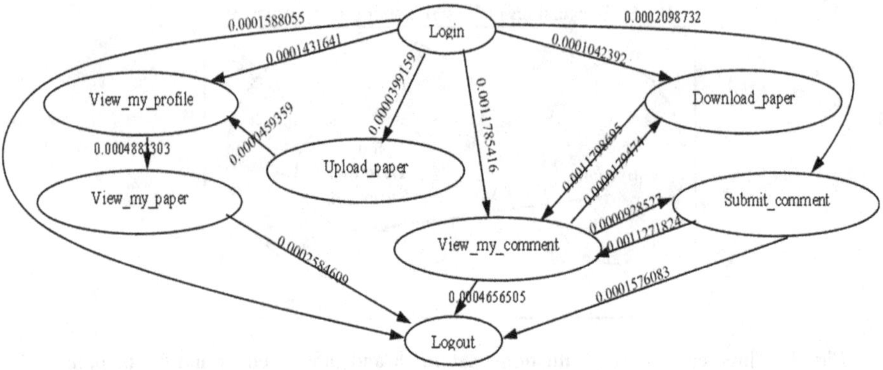

Fig. 4. Part of the situation-transition structure of CoRE dataset

information. Moreover, we have identified 18 end-user desires such as login, view user profile, upload a paper, download a paper, and submit a comment. First, we have derived the situation-transition structure from the observational data of all 65 users and used it to elicit set of new requirements using the procedure given in [3]. Figure 4 shows a part of this situation-transition structure. Each edge is annotated with the mutual information value of the transition. Table 1 listed five sample requirements derived using this part of the structure. These five requirements were unknown when the exiting version of CoRE website was first developed.

Table 1. Sample set of requirements elicited from situation-transition structure generated by considering the behavior of all end-users of CoRE

Requirement	Related situation-transitions
(A) Display the user profile on home page	Login -> View_my_profile
(B) Display the user profile and the list of papers uploaded by the user, once the user uploads a paper	Upload_paper -> View_my_profile -> View_my_paper
(C) Display a link to view submitted comments in download page	Download_paper -> View_my_comment
(D) Display a link to logout, once the user submits a comment	Submit_comment -> Logout
(E) Display the submitted comments once the user submits a comment and then display the link to logout	Submit_comment -> View_my_comment -> Logout

5.1 Evaluating the Individual Requirement Importance Factor

The Eq. (3) is used to evaluate the Importance factor of each individual requirement. Table 2 shows the individual requirement Importance factor of each requirement listed in Table 1.

Table 2. Evaluated individual importance factors of five requirements in Table 1

Requirement	Individual importance factor
(A)	0.0001431641
(B)	0.0005342662
(C)	0.0011798695
(D)	0.0001576083
(E)	0.0015928329

5.2 Evaluating the Risk Factor

In CoRE it is nearly impossible that any end-user behavior could lead to a situation with a physically or mentally harmful or life-threatening outcome. However, end-user behavior in CoRE may cause situations with undesirable outcomes such as uploading or downloading incorrect files, submitting incorrect comments, removing files and comments, etc. Note that the relative risk of each of these outcomes is different. For example, the risk of uploading an incorrect file is higher than submitting incorrect comments in CoRE.

As the first step of evaluating the Risk factors we have labeled each situation in the CoRE situation-transition structure as either risk, safe, safe defender or un-labeled and indicate the relative risk of each situation using a scalar within the range −5 to +5. Table 3 shows the assigned labels and the scalars of the situations related to the requirements given in Table 1. Table 4 shows the evaluated the Risk factors of requirement using Eq. (4).

Table 3. Labeled situations with scalar

Situation	Label	Scalar
Login	Safe	0
Logout	Un labeled	-
View_my_profile	Safe	0
Upload_paper	Risk	4
Download_paper	Risk	1
View_my_comment	Safe defend	−2
View_my_paper	Safe defend	−2
Submit_comment	Risk	2

Table 4. Evaluated risk factors of five requirements in Table 1

Requirement	Risk factors
(A)	0
(B)	2
(C)	1
(D)	2
(E)	0

5.3 Prioritizing the Requirements

The priorities of the requirements are evaluated based on their Importance factors and the Risk factors as given in Eq. (5). Table 5 shows the evaluated priorities of the five requirements given in Table 5. Based on the evaluated priorities the five requirements can be ordered as follows: (E) > (C) > (B) > (A) > (D).

Table 5. Priority of five requirements when $k = 10^4$

Requirement R	$\sim$ Priority(R)
(A)	1.43
(B)	2.67
(C)	11.80
(D)	0.79
(E)	15.93

5.4 Prioritizing the Requirements Based on Individual End-User

As mentioned earlier, the preference of requirements is different for each individual end-user. In order to illustrate this difference we have selected three end-users and derived three separate situation-transition structures by only considering observational data related to each of them. Next, the Importance factors and the Risk factors of the five requirements in Table 1 are evaluated based on these three situation-transition structures. Table 6 shows the priorities of the five requirements with respect to the three end-users. The zero priority implies the absence of situation-transitions related to the particular requirement in the user's situation-transition structures.

Table 6. Priority of five requirements with respect to individual end-user when $k = 10^4$

Requirement R	$\sim$ Priority(R)		
	User 1	User 2	User 3
(A)	15.28	0.00	14.78
(B)	0.00	0.00	65.52
(C)	0.00	20.44	0.00
(D)	18.94	0.00	0.00
(E)	76.30	1.30	128.37

Based on evaluated priorities, the five requirements can be ordered with respect to each end-user as follows:

User 1: (E)>(D)>(A)>(B),(C)
User 2: (C)>(E)>(B),(D),(E)
User 3: (E)>(B)>(A)>(C),(D)

6 Discussion

The proposed requirements prioritization approach provides knowledge on what each requirement means to a particular individual end-user or particular end-user group. We believe this will be useful resource to requirements analysts to make better decisions and to gain higher customer satisfaction on the software.

Moreover, the method can also be used to find the alternative requirements in order to select the most important and low risk requirements to perform a particular task. For example, when comparing the two requirements: (D) allowing end-user to directly logout once the comments are submitted and (E) displaying the submitted comments to the end-user before allowing to logout, the latter has higher priority since the risk is low, although the final results of both requirements are same. In addition, the proposed requirements prioritization method can be performed simultaneously with the requirements elicitation process, and the computationally rich nature of the method makes it possible to automate a significant part of the process. This approach will reduce the time and workload of requirements analysts in finding the final prioritized set of requirements.

Despite these significant benefits, the proposed methodology is not without limitations and potential caveats. Although the total number of situations in a particular domain is finite, the manual process of labeling situations in evaluating the Risk factor of requirements is ineffective or sometime impossible in large complex domains. Clustering of situations into risk groups similar to the requirements clustering approach proposed in [10] would be a promising solution for this problem. However, the feature identification for such machine learning based clustering of situations is challenging. We plan to extensively study the applicability of common features such as term similarity [10] in order to perform such automated clustering of situations based on their potential risk.

The proposed method is highly dependent on observational data and it is impossible to guarantee that the data includes all possible situations and situation-transitions that could exist in a particular domain. The unobserved situations and situation-transitions may affect finding the priorities of the derived requirement. Specially, unobserved risk situations can increase the uncertainty of the evaluated priorities. For example, the zero priority values in Table 6 indicate the unobserved situations and situation-transitions that resulted in un-prioritized requirements for each user. One possible solution would be allowing the requirements analyst to identify the unobserved situations or situation-transitions through the background information and modify the existing situation-transition structure by including nodes for new situations and edges for new situation-transitions. The requirements analysts must also estimate the mutual information of these transitions based on their likelihood. However, this manual approach will significantly reduce the effectiveness of the proposed approach. A systematic procedure of handing the unobserved situations and the situation-transitions will be included in future work.

7 Conclusion

In this paper we present a methodology of prioritizing requirements based on end-users' human factors which is novel to the prevalent methods based on business and system perspectives. Our major contribution in this paper is the introduction of a computationally rich procedure for prioritizing requirements by considering end-users' individual interests and behavioral patterns. As mentioned earlier, the main goal of this proposed approach is to support the existing requirements prioritization approaches in order to provide a better understanding of the end-users' personal interest without direct interaction with users. Moreover, this paper establishes our road map towards introducing a new situation-oriented domain-specific risk analyzing procedures in support of requirements prioritizations, which will be an intriguing key for safety critical systems such as healthcare monitoring, mission critical, security, remote sensing systems, etc. Our future work will also investigate an effective and efficient approach to dealing with unobserved situations and situation-transitions.

References

1. Berander, P., Andrews, A.: Requirements prioritization. In: Aurum, A., Wohlin, C. (eds.) Engineering and Managing Software Requirements, Part 1, pp. 69–94. Springer, Heidelberg (2005)
2. Chang, C.K.: Situation analytics: a foundation for a new software engineering paradigm. Computer **49**(1), 24–33 (2016)
3. Atukorala, N.L., Chang, C.K., Oyama, K.: Situation-oriented requirements elicitation. In: Proceedings of the 2016 IEEE Computer Society International Conference on Computers, Software & Applications (COMPSAC), pp. 233–238 (2016)
4. Brackett, J.W.: Software Engineering Institute Curriculum Module (SEI-CM) 19–1.2. Carnegie Mellon University, Pittsburgh (1990)
5. Leffingwell, D., Widrig, D.: Managing Software Requirements – A Unified Approach. Addison Wesley, Addison Wesley (1999)
6. Mead, N.: Requirements prioritization introduction. Software Engineering Institute Web Publication, Carnegie Mellon University, Pittsburgh (2006)
7. Berander, P.: Using Students as subjects in requirements prioritization. In: Proceedings of the International Symposium on Empirical Software Engineering (ISESE), pp. 167–176 (2004)
8. Bebensee, T., Weerd, I., Brinkkemper, S.: Binary priority list for prioritizing software requirements. In: Wieringa, R., Persson, A. (eds.) REFSQ 2010. LNCS, vol. 6182, pp. 67–78. Springer, Heidelberg (2010). doi:10.1007/978-3-642-14192-8_8
9. Beg, R., Abbas, Q., Verma, R.P.: An approach for requirement prioritization using B-Tree. In: Proceedings of the First International Conference on Emerging Trends in Engineering and Technology, pp. 1216–1221 (2008)
10. Duan, C., Laurent, P., Cleland-Huang, J., Kwiatkowski, C.: Towards automated requirements prioritization and triage. Requirements Eng. **14**(2), 73–89 (2009)
11. Tonella, P., Susi, A., Palma, F.: Interactive requirements prioritization using a genetic algorithm. Inf. Softw. Technol. **55**(1), 173–187 (2013)

12. Perini, A., Susi, A., Avesani, P.: A machine learning approach to software requirements prioritization. IEEE Trans. Softw. Eng. **39**(4), 445–461 (2013)
13. Babar, M.I., Ghazali, M., Jawawi, D., et al.: PHandler: an expert system for a scalable software requirements prioritization process. Knowl. Based Syst. **84**, 179–202 (2015)
14. Glymour, C., Cooper, G.F.: Computation, Causation, and Discovery. MIT Press, Cambridge (1999)
15. Ahmad, A., Shahzad, A., Padmanabhuni, V.K., et al.: Requirements prioritization with respect to geographically distributed stakeholders. In: Proceedings of the IEEE International Conference on Computer Science and Automation Engineering (CSAE), pp. 290–294 (2011)
16. Chang, C.K., Hsin-yi, J., Hua, M., Oyama, K.: Situ: a situation-theoretic approach to context-aware service evolution. Proc. IEEE Trans. Serv. Comput. **2**, 261–275 (2009)
17. CoRE: Situation Centric Intention Driven Research Experiment for Requirements Engineering and Intrusion Detection. IRB ID 14-347. Iowa State University

Requirements Prioritization Decision Rule Improvement for Software Product Line Evolution

Mari Inoki[1(✉)] and Takayuki Kitagawa[2]

[1] Faculty of Informatics, Kogakuin University, Tokyo, Japan
m_inoki@cc.kogakuin.ac.jp
[2] Toshiba Solutions Corporation, Tokyo, Japan
kitagawa.takayuki@toshiba-sol.co.jp

Abstract. In this paper, we propose a method for managing requirements prioritization decision rules. The method was defined on the basis of our experience where the authors participated as project members; the project developed in-house software development support tools based on a software product line. The method consists of a meta-model and processes for utilizing the meta-model. We analyzed the evolution of core assets in relation to an actual project. Tacit knowledge for prioritizing requirements was extracted, made explicit, and used to define decision rules and processes for applying the rules. These rules and processes were specific to the project. To develop decision rules and processes for applying them in other projects, we generalized the basic concept and defined a method for launching, applying and improving requirements prioritization decision rules.

Keywords: Requirements definition · Requirements prioritization · Decision rules · Software product line · Core assets · Software evolution

1 Introduction

Software product line development technology has been an emerging paradigm for developing a software product continuously at lower cost with higher quality in a shorter time in order to meet customer requirements or strengthen an competitive market position [1,2]. In a software product line paradigm, an organization constructs a product roadmap, prepares software assets (core assets), and then develops a software product by reusing the core assets. The market, technology, and organization related to a product line change with time, and the requirements for an existing product line—e.g., removing defects or extending core assets—accumulate. A product line development project must maintain and optimize a product line by evolving core assets [3].

Requirements definition is an important process whereby stakeholders of a software product discover, review, articulate, understand, and document the product's requirements and its lifecycle [4–6]. A project will not always have

S.-W. Lee and T. Nakatani (Eds.): APRES 2016, CCIS 671, pp. 34–49, 2016.
DOI: 10.1007/978-981-10-3256-1_3

sufficient time, resources, and budget to implement all requirements determined by the stakeholders. Therefore, requirements must be prioritized, selected, and implemented in an optimized manner.

In software product line development, projects develop core assets, and then develop a software product by reusing the core assets. The core assets evolve based on development feedback. Similar to single software product development, requirements definition is also important for software product line development. Numerous studies have examined how to prioritize requirements effectively [5, 7–9]. However, little is known regarding requirement prioritization for evolving core assets. The concerned stakeholders change over many years during the evolution of core assets. The basic concept of a product line as defined by the original stakeholders must be maintained, even if the stakeholders change. The requirements for the next model of the core assets must not be prioritized in a self-serving manner; a standardized method for prioritizing requirements is required.

Some requirements prioritization methods that focus on software product line engineering have been proposed [10–12]. However, these are general methods that do not depend on specific domains. To prioritize requirements for the evolution of core assets, concrete knowledge regarding what requirements are appropriate for the next model is required. In addition, determining the relative importance of these requirements is necessary.

We analyzed an actual core asset evolution project, and extracted tacit knowledge for prioritizing requirements. We made such knowledge explicit and defined decision rules and processes for applying the rules. These rules and processes were specific to the project. To apply developed decision rules and processes in other projects, we generalized the basic concept and defined a method for launching, applying and improving requirements prioritization decision rules.

The remainder of this paper is organized as follows. Section 2 analyzes an actual project relative to core asset evolution, and describes problems and solution approaches for requirements prioritization for core asset evolution. Section 3 proposes a method for managing requirements prioritization decision rules. Section 4 evaluates the proposed method. Section 5 presents the conclusions of the study.

2 Problem Definition and Solution Approach

The target is a project in which we had participated as project members; the project developed in-house software development support tools based on a software product line. The developers have developed software products by reusing the core assets and delivered products to more than thirty end users. Requirements were also elicited from end users to improve the core assets and products.

2.1 Core Assets Model Relationships

The project has a product road map that describes a plan to extend the core assets and develop products. Fundamentally, a new model of core assets is

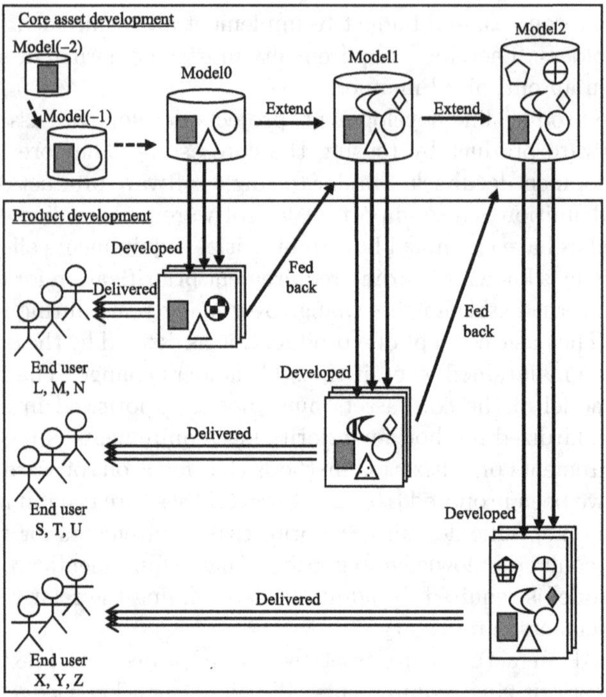

Fig. 1. Relationships between core assets and products

released according to the product road map. The model represents a particular variation of the core assets. The developers have also considered the requirements elicited by the end users. Figure 1 shows the product road map for the target project. There are five core asset models, and we used one of the five models to analyze problems. We refer to this model as Model0. Model1 and Model2 are the succeeding models for Model0 and Model1, respectively. Core assets had been extended twice before Model0. Model(−1) and Model(−2) are the precedent models of Model0 and Model(−1), respectively.

The roadmap did not contain significant architectural changes. Model0 includes Model(−1) and Model(−2); Model1 and Model2 include Model0 and Model1, respectively. Three products for end users L, M, and N were developed by reusing core assets from Model0. The core asset developers considered the requirements elicited from end users L, M, and N, and developed Model1 by extending Model0. Model2 was developed in the same manner as Model1.

2.2 Lead Time for Model0

Table 1 shows the number of requirements and average lead times for defining the requirements. The lead time is the time (in days) between elicitation and agreement for a requirement. The requirements are classified based on the types of requirements and the stakeholders. In this study, we used seven types of requirements; reliability, functionality, usability, portability, efficiency, maintainability, and others. The types except for others were defined based on the quality model [13].

Table 1. Number of requirements and time taken for definition (days) (Model0)

No	Type	Number of req.	Average duration (days)	Number of req.			Average duration (days)		
				End user	Developer	Orderer	End user	Developer	Orderer
1	Reliability	9	145.44	3	5	1	278.00	94.00	5.00
2	Functionality	31	92.77	4	16	11	121.00	114.00	51.64
3	Usability	14	192.29	5	3	6	289.60	80.00	167.33
4	Portability	9	65.33	1	7	1	242.00	48.86	4.00
5	Efficiency	2	4.00	-	1	1	-	0.00	8.00
6	Maintainability	4	154.75	1	3	-	249.00	123.33	-
7	Others	13	54.69	1	4	8	62.00	59.00	51.63
Total		82	107.35	15	39	28	221.27	89.28	124.36

As can be seen in Table 1, 82 requirements were elicited and stored in the requirements list. It took an average of 107.35 days to complete the requirement definitions. Table 1 shows that definition of the reliability, usability, and maintainability requirements took longer than that of the others; the requirements elicited from end users also required more time than the others.

2.3 Moldel0 Analysis

According to observation of Model0, we held discussion with the stakeholders who defined the basic concept of the target product line by using the data of Model0. We encountered the following problems during the requirements definition for a core asset model.

There was no standardized method for prioritizing requirements. Different stakeholders prioritized requirements in their own ways. The definitions of important requirements (including the reliability and usability requirements) tended to be postponed. In addition, even for requirements not included in the product road map, a significant amount of time was required to determine that these requirements were out of the project scope. Consequently, requirements definition took a long time. The extended period of time spent on defining requirements resulted in exceeding the budget.

We analyzed the primary cause of this situation as follows. The core asset models evolved through the inheritance of the precedent model. Stakeholders changed as the core assets evolved. There was no standardized method for prioritizing requirements. Different stakeholders prioritized requirements in their own ways. Worse, they tended to postpone the decision on whether the requirements were within the project scope or not.

2.4 Problem Definition and Solution Approach

The first problem in standardizing a method for prioritizing requirements is that it is not clear what knowledge is necessary for effective decision making in requirements prioritization for core asset evolution. It is considered that there is no universal knowledge for effective requirements prioritization for evolving core assets. As first step, we observed an actual project and extracted concrete knowledge regarding prioritizing requirements related to the core asset evolution.

After extracting concrete knowledge for prioritizing requirements, the next problem we encountered was how to define the appropriate representation form with which different stakeholders could understand and utilize the knowledge easily. Stakeholders changed as the core assets evolved. The representation form of the knowledge must be simple and easy to understand. One simple form of knowledge is a rule: a statement of what requirement can, must, or must not be prioritized by the level of importance in a particular situation. We used a form of rule for representing the extracted knowledge.

These rules and processes are specific to a project. The final problem to be solved was to define a method that allowed other core asset evolution projects to prioritize requirements properly. Based on our experience from the extracting specific decision rules, we defined a method for managing decision rules.

3 Method for Managing Requirements Prioritization Decision Rule

This section describes our proposed method for managing requirements prioritization decision rules. As mentioned in Sect. 2.4, we extracted concrete knowledge of an actual project and defined decision rules for prioritizing requirements specific to the project. Then, we defined a method for managing decision rules. The method consists of a meta-model and its utilization processes. Processes for launching and improving decision rules are also described.

3.1 Extracting Concrete Knowledge

We interviewed the stakeholders who defined the basic concept of the target product line. We extracted four problems from them, showing in (P1)–(P4) of Fig. 2. Then, we extracted know-how (K1)–(K5) regarding the problems and specified viewpoints to focus on. In this case, we focused on the reliability and usability requirements elicited from end users, and functional requirements described in the product roadmap. We represented the extracted know-how in six decision rules.

3.2 Decision Rules and Application Processes

The following are six decision rules devised on the basis of the knowledge mentioned in Sect. 3.1. A stakeholder can avoid the problems described in Sect. 3.1 by adhering these rules.

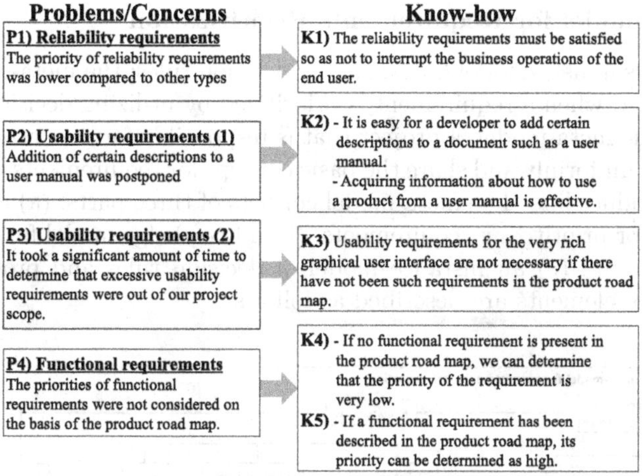

Fig. 2. Problems and extracted know-how

- Rule01: A reliability requirement elicited from an end user must be satisfied. This rule was defined in response to (K1).
- Rule02: A usability requirement elicited from an end user that results in the addition of a particular description to a document must be satisfied. This rule was defined by considering a solution for (K2).
- Rule03: A functional requirement described in the product road map must be satisfied. This rule was defined in response to (K5).
- Rule04: If a requirement is a functional or usability requirement not described in the product road map, the requirement does not need to be satisfied. This rule was defined relative to (K3) and (K4).
- Rule05: If a functional requirement not described in the product road map meets the following conditions, it must be satisfied. This rule was defined as an exception to Rule04.
  (a) It is elicited from an end user.
  (b) The end user is using a software product.
  (c) The end user can make progress with business operations more smoothly with the relevant function.
- Rule06: The priorities of other requirements must be determined individually.

One example application of the rules is in the order Rule01, 02, 03, 05, 04 and 06. Rule05 is applied earlier than Rule04 because Rule05 is an exception to Rule04. Invoking Rule05 corresponds to rescuing a requirement that is essentially regarded as unnecessary. The advantage of using decision rules is that a developer can determine quickly whether a requirement is necessary.

3.3 Meta-model for Requirements Prioritization

Figure 3 shows a meta-model for managing decision rules; the meta-model is the knowledge used when a requirement is prioritized by utilizing decision rules. By referencing the meta-model, a project that is responsible for core asset evolution can represent uniformly and share the basic concept for requirements definition of the target product line. The meta-model consists of three parts: (a) a knowledge meta-model for prioritizing requirements, (b) a knowledge model for prioritizing requirements, and (c) an instance model of decision rules and processes. The features of the elements are described as follows.

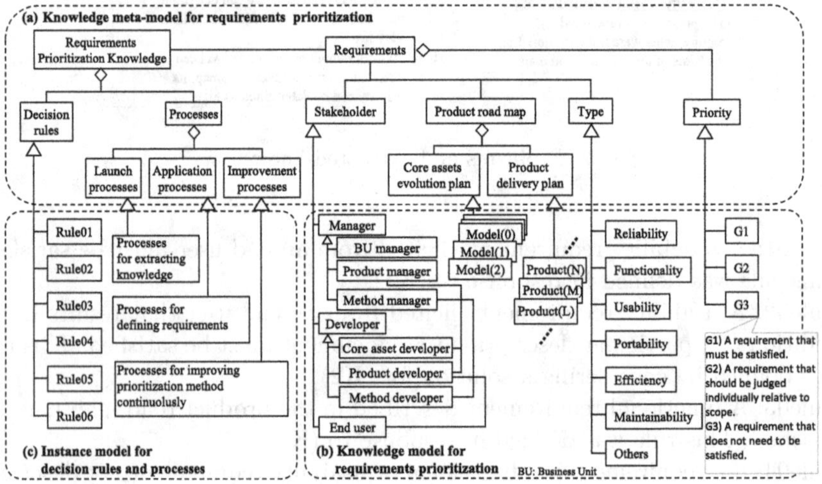

Fig. 3. Meta-model for prioritizing requirements

(a) The knowledge meta-model for prioritizing requirements is a basic concept regarding how to prioritize requirements and manage decision rules. It shows that the priority of a requirement is defined on the basis of decision rules and that the decision rules are managed through the processes for launching, applying and improving them. Basically, the priority of a requirement is determined from the viewpoint of the stakeholder type, product road map, and requirement type. Such decision knowledge is specific to an experienced engineer. Therefore, such decision knowledge is specified and made explicit in the form of a decision rule. A decision rule enables a project member to share and utilize knowledge uniformly.

(b) A knowledge model for prioritizing requirements is a particular example of the right hand part of the meta-model shown in Fig. 3(a). This indicates instances of a stakeholder type, core asset evolution plan, requirement type, and priority.

The model shown in Fig. 3(b) is the knowledge extracted from a previous project; it is not applicable to every project. However, referencing the model enables us to define the knowledge model of another project easily.

Priority instances are classified on the following three levels.

(G1) A requirement that must be satisfied.
(G2) A requirement that should be judged individually relative to its scope.
(G3) A requirement that does not need to be satisfied.

Priority levels G1, G2, and G3 are presented in order of priority. The priority of G2 requirements should be determined depending on the individual situation. Developers must investigate individual requirements. G3 requirements do not need to be satisfied, i.e., they are beyond the scope of the target product line.

(c) An instance model of decision rules and processes is a particular example of the left hand part of Fig. 3(a). It is also an example of the knowledge for prioritizing requirements. When a project that is responsible for core asset evolution begins, a project member should define processes for launching, applying and improving their own decision rules; following such processes, decision rules should be launched, applied and improved.

The model shown in Fig. 3(c) includes six decision rules and their processes that were defined according to the analysis of the previous project. This instance model depends on such factors as domain type, stakeholders, and the concerns they face.

3.4 Meta-model Utilization Process

The primary stakeholders related to the software product line evolution development are the business unit manager, product manager, developer, and end user. The business unit manager is in charge of decision making for the investment of core asset development on the basis of a business plan.

The product manager manages core asset developers and product developers. The product developers reuse the core assets developed by the core asset developers. The end users utilize the products delivered by the product developers. The end user corresponds to both individuals and organizations. Using the method effectively requires a method manager and a method developer who are in charge of managing or developing the method for the core asset evolution.

Figure 4 shows processes for utilizing the meta-model. A method manager launches, monitors and evolves a knowledge meta-model. A product manager draws a product road map and a business unit manager authorizes the road map considering other product roadmaps under his or her business unit. The product road map is monitored and evolved. A method developer defines, monitors and evolves a knowledge model and processes. After that, the method developer defines decision rules on the basis of a knowledge model and processes. The decision rules are monitored and evolved. A core asset developer defines, monitors and evolves core assets; requirements of core assets are defined by applying the decision rules defined by the method developer.

3.5 Launching Process

Based on the meta-model utilization processes, the processes for launching decision rules were defined as follows. The processes for launching requirements

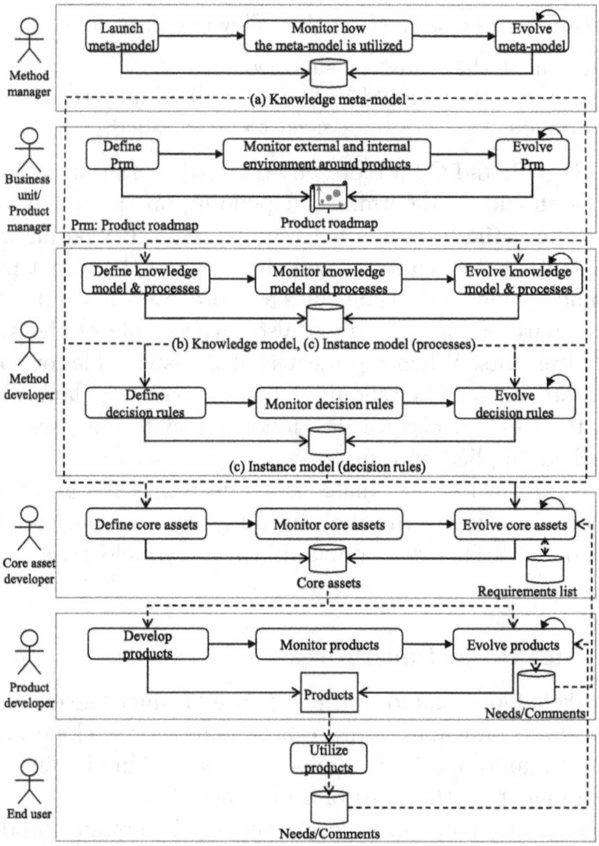

Fig. 4. Processes for utilizing the meta-model

priority decision rules consist of four sub processes. The first sub process is to interview the stakeholders and ask them to point out problems and concerns regarding the results of the past requirements definition. The second sub process is to extract know-how from their answers and specify viewpoints which they focus on. The viewpoints include kinds of requirements, stakeholders, and specific features. After that, the relationships between such viewpoints and requirements priorities should be clarified. The last process is to represent the extracted know-how in the form of a rule on the basis of the related viewpoints and priorities.

3.6 Improvement Process

Here, we define the processes for improving the requirements prioritization method in the same manner as the launching processes. The processes consist of an iteration of the plan-do-check-act (PDCA) cycle. The PDCA cycle is a management tool that assumes that every managerial action can be improved by careful application of the PDCA sequence. An important contribution to continuous process improvement is the kaizen, a movement that leads to continual and incremental improvement [14]. Through this cycle, an organization improves a knowledge model, decision rules, and processes. The multiplier effects of kaizen help an organization maintain optimal methods at all times.

4 Evaluation

4.1 Evaluation Method

We have evaluated the proposed method from two points of view. First, we evaluated the effectiveness of the six decision rules using two models, Model1 and Model2, which were discussed in Sect. 2. By taking Rule01 to Rule06 as the initial rules, each requirement of Model1 was analyzed to determine if these rules are applicable; thus, the priority of the requirement was defined. The core asset developers of Model2 were different from those of Model1. We evaluated the effectiveness of the Model2 rules and application processes under the condition that different stakeholders utilized them. In addition, the effectiveness of the improvement processes was evaluated. Those two models and the current model, Model0, were developed in different fiscal years. However, the duration for requirements definition for all three models was six months.

In the second evaluation, we applied our method to three other projects. In particular in order to evaluate the effectiveness of a meta-model and utilization processes; we focused on the processes for launching decision rules.

4.2 Applied Rules

Table 2 shows the number of decision rules applied to Model1 and Model2. The number of candidate requirements for Model1 and Model2 were 20 and 27, respectively. These candidate requirements were classified into the requirements G1, G2, or G3 shown in Table 2. The requirements classified as G2 by application of Rule06 were analyzed. These requirements were classified as G1 or G3 requirements. Since the scope of some requirements was small, it was difficult to judge whether the product road map included them.

Table 2. Rules applied to Model1 and Model2

	Rule01	Rule02	Rule03	Rule04	Rule05	Rule06		Total
	G1	G1	G1	G3	G1	G2		
						G1	G3	
Model1	0	9	2	1	0	7	1	20
Model2	7	4	5	0	2	6	3	27
Total	7	13	7	1	2	13	4	47
Ratio	14.89%	27.66%	14.89%	2.13%	4.26%	27.66%	8.51%	-

Although Rule05 was not applied to any requirements for Model1, we used the decision rules as they were for Model2. According to Table 2, all of the rules were utilized, though the utilization ratio of Rule06 was the highest. It should be acknowledged that additional rules could be added.

4.3 Duration for Requirements Definition

(1) Duration for the requirements definition of Model1 and Model2: The results of applying the requirements prioritization method to Model1 are shown in Table 3. The results for Model2 are shown in Table 4. As can be seen, an average of 1.15 and 15.19 days was required to define the requirements for Model1 and Model2, respectively.

(2) Solution Situation for the Long Duration Problem: According to the information shown in Table 3 and Table 4, an average of 1.5 days or less was required to define the reliability requirements of Model1 and Model2. Seven requirements were determined by focusing on the reliability requirements of Model2, which were elicited from end users. Defining these requirements required an average of 1.71 days. Compared to Model0, the duration of the reliability requirements definition for Model1 and Model2 appears to be improved.

Table 3. Number of requirements and time taken for definition (days) (Moldel1)

No	Type	Number of req	Average duration (days)	Number of req.			Average duration (days)		
				End user	Developer	Orderer	End user	Developer	Orderer
1	Reliability	2	0.50	0	1	1	-	0.00	1.00
2	Functionality	4	1.25	2	1	1	0.50	3.00	1.00
3	Usability	10	1.40	9	0	1	1.44	-	1.00
4	Portability	1	1.00	0	0	1	-	-	1.00
5	Efficiency	0	-	0	0	0	-	-	-
6	Maintainability	1	0.00	0	1	0	-	0.00	-
7	Others	2	1.00	0	0	2	-	-	1.00
Total		20	1.15	11	3	6	1.27	1.00	1.00

The definition of usability requirements for Model1 and Model2 required 1.4 and 19.38 days, respectively, on average. Model2 required more time than Model1 owing to usability requirements that were difficult for developers to understand. The developers needed to interview end users regarding these requirements. Compared with the duration of the usability requirements for Model0, both Model1 and Model2 appeared to improve.

Regarding the definition of functional requirements, Model1 and Model2 required 1.25 and 6.5 days, respectively, on average. The long duration problem of the functional requirements definition was improved compared to the 92.77 days required for Model0.

Table 4. Number of requirements and time taken for definition (days) (Model2)

No	Type	Number of req	Average duration (days)	Number of req.			Average duration (days)		
				End user	Developer	Orderer	End user	Developer	Orderer
1	Reliability	8	1.50	7	0	1	1.71	-	0.00
2	Functionality	8	6.50	8	0	0	6.50	-	-
3	Usability	8	19.38	6	1	1	25.83	0.00	0.00
4	Portability	1	0.00	1	0	0	0.00	-	-
5	Efficiency	0	-	0	0	0	-	-	-
6	Maintainability	0	-	0	0	0	-	-	-
7	Others	2	191.00	1	1	0	0.00	191.00	-
Total		27	15.19	23	2	2	779.52	95.50	0.00

Four requirements were classified as others in Tables 3 and 4. These were the requirements for project management. For example, they concerned the improvement of staff assignment and resource distribution. There was one requirement that required 191 days to define for Model2. This requirement was related to an in-house project management standard. The long definition period originated in project management and was unrelated to the requirements prioritization method.

4.4 Cost Analysis

Table 5 shows the costs to define requirements for Model0, Model1, and Model2. Cost was measured by taking a requirement from each requirement list for Model0, Model1, and Model2. The value of the cost is relative and does not have a specific unit. As is shown in Table 5, the costs of requirements definition for Model1 and Model2 were reduced compared to Model0. The periods for defining requirements for Model1 and Model2 were shorter than that for Model0 because it was not necessary to repeat requirement understanding, prioritizing, and reviewing activities. This reduced the costs of the requirements definition for Model1 and Model2.

Table 5. Cost for defining one requirement

	Model0	Model1	Model2
Elicitation	1.25	0.5	0.7
Prioritization	1	0.42	1
Review & Agreement	0.75	0.42	0.75
Total	3	1.34	2.45

Table 6. Cost for defining 20 requirements

	Model0	Model1	Model2
Elicitation	25	1	11
Prioritization	20	2	16
Review & Agreement	15	9	11
Total	60	12	38

Twenty requirements were determined for Model1. Since the developers could concentrate on the requirements definition in a short time, the definition of these 20 requirements was performed simultaneously. Table 6 shows the costs to define the 20 requirements. The 20 requirements definition costs for Model0 and Model2 were calculated by multiplying 20 by an individual cost for the requirements shown in Table 5. According to Table 6, the costs for Model1 and Model2 were reduced to 20% and 63% of Model0, respectively.

4.5 Improvement of Rules

There is a portability requirement when Rule06 is applied for both Models 1 and 2. A portability requirement for the core assets evolution indicates that a software product developed on the basis of a core assets model can work for a new core assets model. There are several solutions to address the implementation of a portability requirement. One solution is to prepare an automatic transformation mechanism. Another is to provide a document that describes the procedure for converting the product to a new environment. End users must be able to utilize a new core assets model with one of the solutions. The portability requirements must be determined immediately as within the scope of the next model.

In the evaluation, we did not prepare official decision rules related to portability requirements. However, the core assets developers tacitly placed high priority on portability requirements, thereby obviating any need to have a significant period for defining requirements for the portability requirement.

4.6 Launching Decision Rules for Other Projects

We applied our method on the following actual projects. In particular, we focused on the processes for launching decision rules.

(1) Embedded software: An organization planned to produce software products for not only the Japanese market but also global markets. The organization planned to extend the core assets and prepare core asset variations regarding the graphical user interface, voice message, and regulations of individual countries. Though the organization maintains core assets for several years, some problems occurred. For example, there are core assets that are dormant. Core assets including unnecessary ones entail cost for product validation. Core assets that are similar to each other prevent a product developer from progressing.

It is important for the production of products for global markets to extend core assets. However, before extending the core assets, it is more effective to perform the refactoring of the core assets and to improve their maintainability in order to increase the variations. In accordance with this knowledge, we devised the following rule: *a maintainability requirement for defining clear variation points for globalization must be satisfied before extending the core assets.*

(2) Media management support system: A media management support system helps an editor edit several types of data, including text, graphic, sound, and video. Such a system is designed for an organization of a specific business domain. An end user is required to extend and connect several existing systems together to the media management support system. This allows the end user to gather data easily through the media management support system.

There are several kinds of organizations. Some organization are small and have few related systems that need to be connected. Others are very large and need to connect several existing systems together. Therefore, the business unit which has maintained the core assets for the media management support system is planned to prepare interfaces to connect existing systems as elements of core assets. However, there are several types of existing systems among the target organizations; it is difficult to prepare interfaces which adopt several types of existing systems in advance. In addition, preparing many core assets entails a high cost. Instead of developing several interfaces that might not be used, it is effective to specify an instruction document and share the document as a core asset; the document includes features of an existing system, the reason the customer needs to connect to the media management system, the interface specification of the systems, and how to develop the interface. In other words, the business unit realized that if it is difficult to prepare several elements of core assets in advance, it is effective to prepare an instruction document for developing such elements as another core assets.

According to the above knowledge, we extracted the following rule: *preparing a document for developing elements of core assets takes priority over developing the elements themselves.*

(3) Product information traceability system: Product information traceability is the ability of the product information to be traced along the audit workflow. Product information includes the location, date, and status of products and claims from stakeholders. This system processes such information. End users refer to the product information accumulated in the system and utilize it for checking the safety of the products. Basically, there are two kinds of product model: a model for large order-made products and a model for small mass-produced items. The former traces products individually, and the latter traces the product on the basis of the level of importance of the information. This system is utilized by the end-users in order to obey the laws and regulations regarding the target products. Therefore, the requirements for verifying and showing evidence of the correctness of the audit processes on the basis of the laws and regulations and for changes in laws and regulations are necessary.

According to the knowledge above, we defined the following rule: *a requirement alternation based on the changes in laws and regulations regarding the workflow processes must be satisfied.*

5 Conclusion

In this paper, we proposed a method for managing requirements prioritization decision rules and reported an application of the method. The method was defined for a core asset evolution project. The objective of the method is to help different stakeholders define the priority of a requirement uniformly. We defined decision rules for prioritizing requirements and processes for applying the rules. In addition, we defined a meta-model for prioritizing requirements and processes for continuous improvement of rules. According to an evaluation of the method, it was clarified that different stakeholders were able to reach agreement smoothly and efficiently. In addition, the cost required to define requirements was reduced. Our requirements prioritization method helps a core asset evolution project continuously acquire and share concrete knowledge regarding what requirements are appropriate for the next model. This incremental PDCA cycle plays the role of an engine for an organization's growth. As future work, we will increase the number of projects, and we will contribute to a business's growth.

References

1. Clements, P., Northrop, L.: Software Product Lines: Practice and Patterns. Addison-Wesley, Boston (2001)
2. McGregor, J., Muthig, D., Yoshimura, K., Jensen, P.: Guest editor's introduction: successful software product line practices. IEEE Softw. **27**, 16–21 (2010)
3. Botterwec, G., Pleuss, A.: Evolution of software product lines, Chap. 9. In: Mens, T., Serebrenik, A., Cleve, A. (eds.) Evolving Software Systems. Springer (2014)
4. IEEE std. 830–1998: IEEE Recommended Practice for Software Requirements Specifications. IEEE (1998)
5. International Institute of Business Analysis: A Guide to the Business Analysis Body of Knowledge® (BABOK® Guide), Version 2.0 (2009)
6. ISO, IEC, IEEE: ISO/IEC/IEEE 29148, Systems and Software Engineering, Life Cycle Processes, Requirements Engineering (2011)
7. Karlsson, J., Wohlin, C., Regnell, B.: An evaluation of methods for prioritizing software requirements. Inf. Softw. Technol. **39**, 939–947 (1998)
8. Mead, N.R.: Requirements Prioritization Introduction, Carnegie Mellon University, (2006–2013) https://buildsecurityin.us-cert.gov/articles/best-practices/requirements-engineering/requirements-prioritization-introduction
9. Devulapalli, S., Khare, A., Rao, O.R.S.: Requirements prioritization: survey and analysis. In: Satapathy, S.C., Bhatt, Y.C., Joshi, A., Mishra, D.K. (eds.) ICICT 2015. AISC, pp. 567–575. Springer, Heidelberg (2016)
10. Lee, J., Kang, K.C., Sawyer, P., Lee, H.: A holistic approach to feature modeling for product line requirements engineering. Requirements Eng. **19**, 377–395 (2013). Springer

11. Derakhshanmanesh, M., Fox, J., Ebert, J.: Requirements-driven incremental adoption of variability management techniques and tools: an industrial experience report. Requirements Eng. **19**, 333–354 (2013). Springer
12. Alferez, M., Bonifacio, R., Teixeira, L., Accioly, P., Kulesza, U., Moreira, A., Araujo, J., Borba, P.: Evaluating scenario-based SPL requirements approaches: the case for modularity, stability and expressiveness. Requirements Eng. **19**, 355–376 (2013). Springer
13. ISO, IEC: ISO/IEC 9126–1: 2001, Software Engineering - Product Quality - Part 1: Quality Model (2001)
14. Imai, M.: Kaizen - The Key to Japan's Competitive Success. McGraw-Hill, New York (1986)

Requirements Modeling and Process for Quality

Incorporating Sustainability Design in Requirements Engineering Process: A Preliminary Study

Theresia Ratih Dewi Saputri[1] and Seok-Won Lee[2(✉)]

[1] Department of Computer Engineering, Ajou University, Suwon, Korea
trdsaputri@ajou.ac.kr
[2] Department of Software and Computer Engineering,
Ajou University, Suwon, Korea
leesw@ajou.ac.kr

Abstract. Sustainability is usually treated as an after-thought in software engineering practice. The software engineers tend to focus more on the technical dimension rather than the entire sustainability dimension. Designing software sustainability is a big challenge in current software engineering practices due to the lack of well-defined guidelines that provide tangible decomposition of sustainability aspect. Thus, we propose a framework to analyze the sustainability dimension and structure it into software requirements. Moreover, the research goal of this paper is to develop a methodology that determine sustainability requirements. The proposed meta model integrates four sustainability dimensions with the other quality attributes such as performance and usability. The contribution of this work is to help the requirements engineer to incorporate sustainability concern into software system design.

Keywords: Software · Sustainability · Requirements · Goal modeling · Meta model

1 Introduction

Nowadays, software system becomes a part of the human life. It has been used in the various aspect in the society such as healthcare, education, and entertainment aspect. The current society is facing challenge in sustainability that requires long-term and integrated thinking. Software Engineering field plays important role in the sustainability in the perspective of software as a part of human life. The advancement of technology in the software systems affect the society and environments. Some studies [1–3] have been conducted to examine the global impact of software system. One of the major impacts that have been reported is the energy consumption and hardware disposal that result in the increase of carbon footprint.

Bruntland Commission of United Nations in [4] defines sustainability development as "development that meets the needs of the present without compromising the ability of future generation to meet their own needs". The current practices of software engineering lack of the long-term thinking which is the core component of sustainability development. The consideration beyond immediate software product quality is

S.-W. Lee and T. Nakatani (Eds.): APRES 2016, CCIS 671, pp. 53–67, 2016.
DOI: 10.1007/978-981-10-3256-1_4

usually treated as the secondary concern. The software engineers tend to focus more on the technical dimension instead of the entire sustainability dimensions. Even though some researchers argue that technical dimension is the main concern of the engineer in term of sustainability development, it is strongly dependent with its organizational and business context. Therefore, we cannot think that technical dimension as separate entity from social and financial dimension. The practice to incorporate sustainability concern into the design of software remains missing.

Due to the complex concept of sustainability, designing a software system with the sustainability criteria is not a trivial task. Addressing sustainability should be done in the early phase of SE. The approach to discover sustainability requirements has been studied in the past few years. However, the existing works mostly discussed method in the conceptual level. There is no practical guideline to determine the sustainable requirements of a software system. Without a well-defined guideline, sustainability becomes hard to measure. This paper proposes an approach to decompose the sustainability requirements. Moreover, the goal is to develop a methodology that determine sustainability requirements from the entire sustainability dimension. The contribution of this paper is a framework for modeling sustainability requirements based on stakeholders needs and business context. The framework is intended to help the requirements engineer to incorporate sustainability concern in the software system design.

The remainder of this paper is as follows: Sect. 2 discusses the motivation and the existing works addressed sustainability in software engineering, especially for capturing sustainability requirements. Section 3 presents the proposed approach for modeling sustainability. Section 4 discusses the result of the work by applying the proposed approach in a particular case study. Section 5 states the treats of validity in this research including construct validity, internal validity, external validity and reliability. Then, Sect. 6 concludes the paper.

2 Background and Related Works

This section first discusses about sustainability in the software engineering field. Then, it also presents the important of addressing sustainability in requirements engineering process.

2.1 Sustainability in Software Engineering

In the past few years, sustainability has been considered in the SE process. The researchers in SE field tried to define sustainability from the perspective of SE. Tate [5] defines Sustainable SE as a development which is able to make a balance between rapid release and long term sustainability. Timo et al. [6] explain the goal of Green and Sustainable Software as "the enhancement of SE which targets the direct and indirect consumption of natural resources and energy as well as the aftermath which are caused by software system during their entire life cycle".

Various studies have been conducted to address sustainability concern in software life cycle. Nauman et al. [7] proposed reference model for green and sustainable SE. Dick et al. [8] presented a life cycle thinking inspired life cycle for software products.

Even though, those works showed remarkable process to incorporate sustainability the entire life cycle of SE, those works focus in a conceptual level. There is no discussion about the different dimension of sustainability.

It is important to take different dimensions of sustainability into account. Sustainability should be considered as an integrated concept. Becker et al. [9] proposed a manifesto for sustainability called The Karlskrona Manifesto. It discussed the software design with sustainability concern by taking five sustainability dimensions into account. By referring to Karlskrona Manifesto, the proposed work in this paper also considered different dimension in sustainability development.

2.2 Capturing Sustainability Requirements

One of the most important processes in SE process is requirements engineering (RE). The foundation of software system development strongly dependents on the RE process. Therefore, considering sustainability aspects should be done in the RE phase. Becker et al. [10] argued that RE has important task in sustainability by understanding the nature of software systems and their impacts in the entire dimension of sustainability development.

Even though sustainability becomes a major concern in the society, considering sustainability design in SE is not a trivial task. Chitchyan et al. [11] discussed some of the challenge in considering sustainability design. One of the significant challenge is the lack of methodological support. They argued that there is a limited methodological support for practicing sustainability design. Some works have been proposed to incorporate sustainability design in RE. Mahaux et al. [12] presented an assessment of the software project which take sustainability as one of the quality requirements. The proposed method only focused on the environmental dimension of sustainability. The work proposed by Penzenstadler and Femmer [13] aims to integrate sustainability in software development by incorporating the entire sustainability dimensions. They proposed a generic conceptual model for sustainability including a meta model which consists of sustainability value, indicator, and activity.

The modeling technique proposed by Penzenstadler and Femmer [13] showed significant result in analyzing sustainability design in software system. Our proposed approach adopted their meta model to analyze the sustainability property. However, there is no way to assess the potential conflict from the defined sustainability requirements in their proposed approach. Therefore, one of our aims is to extend the meta model by providing a technique to identify the trade-off of the sustainability requirements from different dimensions. We also aim to couple the sustainability design with other software quality attributes.

3 Proposed Approach

This paper proposed a framework that is used to determine the sustainability requirements from stakeholders needs. The requirements are gathered based on four sustainability dimensions which are social, technical, economical, and environmental dimension. Figure 1 shows the proposed sustainability requirements model. In order to

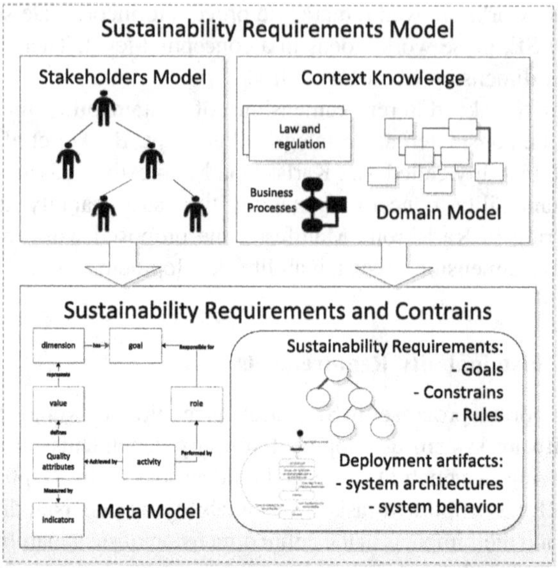

Fig. 1. Artifacts model for modeling sustainability requirements

capture sustainability requirements and constrains, stakeholder model and context knowledge are used as the sources. Those represent the sustainability goal from stakeholders' viewpoints.

There are three main steps in the proposed approach as seen in Fig. 2. The first step is defining sustainability goals. By using Goal-Question-Metric (GQM) approach [14], the sustainability goals are elicited from the sources. The second step is analyzing sustainability properties. In this step the sustainability requirements are determine based on the proposed meta model. The impact and trade-off analysis of the captured requirements is also performed in this process. Then, the last step maps the action from sustainability requirements to runtime model for adaptability.

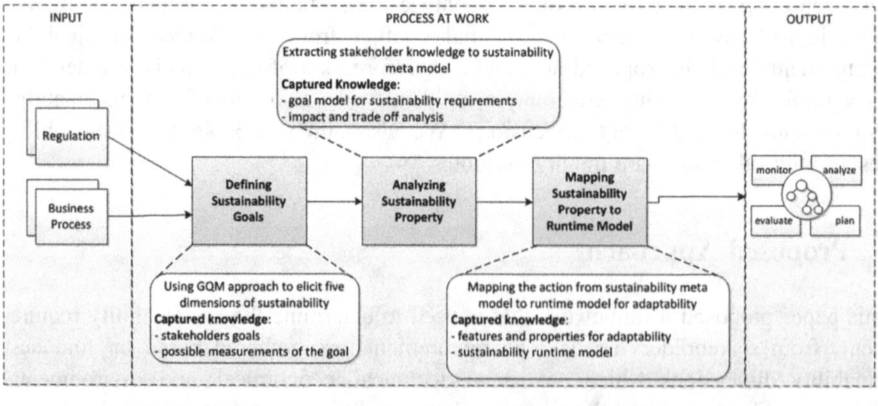

Fig. 2. Proposed approach for modeling sustainability requirements

3.1 Defining Sustainability Goals

In order to elicit stakeholder needs, we use GQM approach along with the defined questions. In order to utilize GQM approach, there are four main process that should be followed which are planning, definition, data collection and interpretation.

The first process to define sustainability goals is collecting the knowledge including stakeholder information, business processes, law and regulation and domain model. The entire stakeholders should be identified. After the stakeholders list has been gathered, the next process is asking question related to each sustainability. In order to capture the goal and metric for the sustainability requirements, 5W + 1 H questions should be asked. Table 1 shows how the defined question can be used to capture the stakeholders needs.

Table 1. Guidance question to identify stakeholders needs

Type	Purpose	Question(s)
who	goal	who is the source? who is the responsible actor?
what	goal	what is the sustainability goal? what is the feature?
which	goal	which sustainability dimension we are satisfying?
when	goal	when it should be run?
where	goal	where it should be located?
how	metric	how can it be tested?

3.2 Structuring Sustainability Requirements

As mentioned in the previous section, the meta model proposed in [13] had been adopted and extended in this work. It presents remarkable meta model to decompose sustainability and relate it to software system development. We adopted some elements from the proposed meta model such as Dimension, Value, Indicators, and Activity. In order to improve the meta-model, we added the Sustainability Dimension, Quality Attributes and Role elements as seen in Fig. 3.

Sustainability dimension is the essential elements in the meta model. This element is used to identify the sustainability requirements in each dimension of sustainability which include social, environmental, technical and economic dimension. The other element which is Quality Attributes element is the quality requirements related to a particular sustainability goal. The Role element is used to represent the actor that has responsibility to perform the activity. It is important to couple sustainability with the other quality attributes. By considering sustainability as quality requirements we are able to think about developing sustainable software rather than sustaining existing software. Another important element in the meta model is Indicators. This element is used to evaluate the achievement of the requirements. Therefore, it should be aligned with the sustainability Goal element.

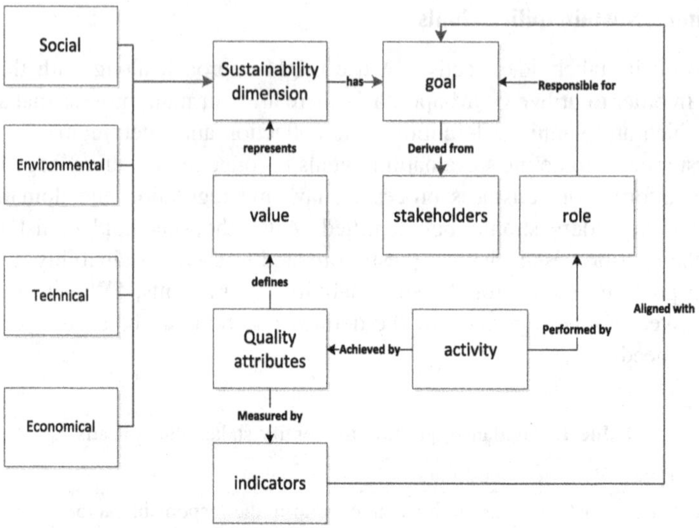

Fig. 3. Meta model for structuring sustainability goal

3.3 Sustainability Requirements Trade-Off Analysis

In order to defined sustainability requirements, we adopted the four sustainability dimensions in [9] and relate them with the other software system quality attributes. The goal determined from a certain dimension may influence the goal from another dimension. For example, in order to increase the performance in technical dimension, we need to increase the cost for purchasing a better hardware. Knowing that there is a possible conflict among the sustainability goals, we also propose a trade-off analysis template that is used to enhance the process for identifying the influence and priority of the requirements.

Table 2 shows the proposed trade-off analysis template that can help the engineering to design the software system. T, En, S, Ec represent technical, environment, social, and economical dimension respectively. In this process, the analysis will be asked to determine the influence of the entire n-th goal (Gn) in each dimension of the sustainability. The influence will be represented by using certain arrow. The up-ward arrow (↑) indicates supports influence. On the other hand, the downward (↓) arrow indicates conflict influence. In addition, the rightward (→) arrow indicates a neutral influence.

Table 2. Goal influence summary

Goal	T	En	S	Ec
G_1	↑	→	→	↑
…	…	…	…	…
G_n	↑	↑	↑	↑

In order to support decision making process, we also provide a method to determine the goal priority as seen in Table 3. The priority assigned on a certain goal will be determined based on the points given by the stakeholder (Sn). Firstly, the stakeholder will be assigned with a certain weight in the range from 1 (lowest) to 100 (highest). Secondly, the point for each goal is determined by the stakeholder in the range from 1 (low priority) to 5 (high priority). Lastly, the point of each goal will be multiplied by the stakeholder's weight as the final point of the goal.

Table 3. Stakeholder goal priority

Stakeholder	Weight	G_1	G_2	…	G_n	Total
S_1						
S_2						
…						
S_n						
Total						

3.4 Runtime Meta Model

One of the most important goal of software sustainability is the ability to rapidly change its behavior to meet the new needs based on environment change. Therefore, this proposed approach also proposed a sustainability runtime model that is used to provide adaptability in software system. The runtime model is used to define the adaptation behavior of the system. There are four entities in the meta model which are condition, state, transition and action as seen in Fig. 4. By using the meta model, we are able to define the trigger for the adaptation and what kind of action that needs to be taken to meet with the change of environments.

In order to perform adaptation, we adopt MAPE-K approach proposed by IBM. As we know about MAPE-K approach, we have monitor that continuously watch the running system and the environment condition. This information will be analyzed by the analyze entity. This entity reasons the environment condition and match it with the

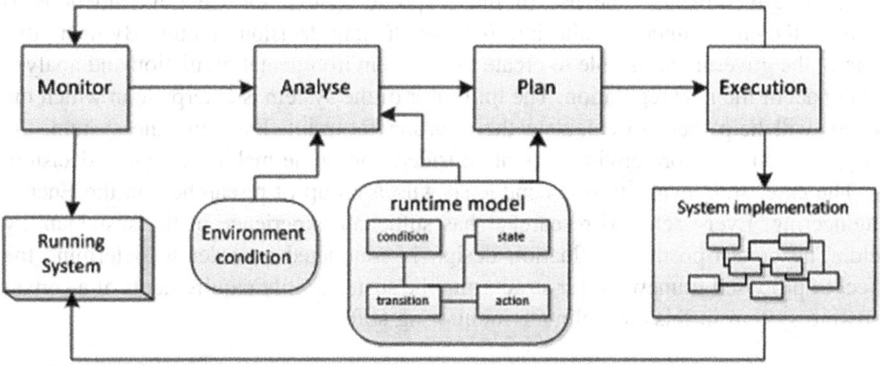

Fig. 4. Sustainability runtime process based on MAPE process

runtime model. When an adaptation action is needed, the plan entity will work on the possible action that should be taken. Once the action is ready, the execution entity will change the system configuration based on the new requirements. Then, the new system implementation will be used as the running system configuration.

4 Result and Analysis

In this section, we discussed the case study that is used to evaluate the implementation of the proposed approach. This section also explained in the process in work for implementing the framework.

4.1 Case Study

In order to show the applicability of the proposed approach, the climate change domain is introduced. The climate change has been one of the emergent global issues in the 21st century. The climate models heavily effect on various areas such as weather events, agriculture, and ecosystem. Therefore, climate modeling becomes complex problem which is hard to understand, predict and solve. In order to address the challenges in climate changes, the researcher from diverse areas should contribute and collaborate.

Easterbrook in [15] listed the role of software engineering community in climate change domain. One of the important roles is the ability of software engineer to analyze and prioritize the multi-stakeholders' requirements. The other role is to build the information system and knowledge management tools that can be used to help the decision making process. Due to those particular reasons, climate change domain is chosen as the case study to evaluate the proposed method.

The climate monitoring system shown in Fig. 5 aims to support effective decision making process. There are three types of user in this system which are individual user, governments and enterprise. The purpose of the system for individual user is to build a media for environment education. By using the system, individual user is able to estimate the impact of their behavior to the environments. The second user in this system is governments. The use of the proposed system for the government is to facilitate the environment monitoring to make efficient decision making. By using this system, the government is able to create the new environmental regulation and analyze the impact of the new regulation. The third user of the system is enterprise in which the system will help them to calculate their carbon footprint. By using the system, the enterprise can get more environmental consideration while making business decision.

The case study evaluation is conducted with a group of researchers in the Energy Engineering. Every selected researcher has sufficient experience in the sustainability field. The pretest-posttest evaluation design is conducted in order to determine the effect of proposed framework for structuring the sustainability requirements of a certain software system in this case climate monitoring software.

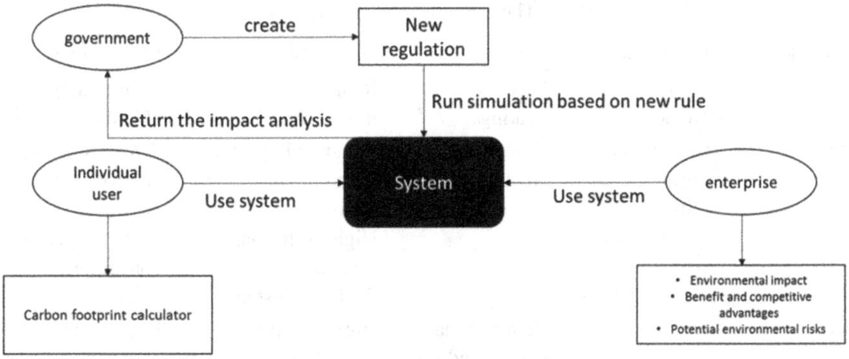

Fig. 5. Conceptual model of climate monitoring system

4.2 Implementation

In order to gather the requirements from the stakeholders, we first identified the possible stakeholder for the climate monitoring system. Due to the complexity of the climate monitoring system, we simplified the case study by choosing some stakeholders for the system. The stakeholders are grouped into three categories which are user, knowledge providers and developers. As mentioned in the previous section, there are three kind of user for the system such as individual users, enterprise and governments. The second category is knowledge providers which consist of the information expert such as geo-scientist and environmentalist. The last category is developers which consists of service provider and software developers.

Following the GQM approach, we then determined the goals of the entire stakeholder. The goals are identified based on four sustainability dimensions which are economic, social, technical, and environmental. Table 4 show some of the example of the stakeholder goals in the entire dimensions of sustainability.

The second process in the proposed method is analyzing the sustainability properties. There are two phases that should be done in the process. The first is constructing the goal model from the sustainability requirements. By using the result of the GQM approach, we analyzed the requirements of the stake-holders in term of the proposed meta model for sustainability requirements. For each stakeholder, the value and activity which are related to the dimension should be decomposed.

Figure 6 shows the example of goal model from Enterprise. For the environment dimension one of the values is Reduce Energy Consumption. In order to achieve the value, some activities such as Reduce Electricity, Reduce Heating, and Find Effective Behavior. Each of the activity will be measured through a certain indicator that have been defined in the previous process. For example, Reduce Electricity Use will be measured based on the Power Bill. During this process, we also need to identify the influence of the sustainability requirements and quality attributes.

As we can see in the Fig. 6, Performance attributes will influence the energy consumption and development cost. However, this goal model does not clearly show what is the influence of the quality attributes. One goal can have positive influence and

Table 4. The gathered stakeholders' goals

Stakeholder	Economic	Social	Technical	Environment
Governments	Cost efficiency	Better rule making	Real time monitoring	Better energy efficiency
Enterprise	Gain higher profit minimize development cost	Competitive advantages	Integrated system with climate factors High performance system Scalable system	Avoid potential environmental risk Reduce energy consumption
User	-	Better climate understanding Knowledge sharing	High usability Fast performance	Energy saving
Developers	Cost efficiency	-	High performance system Scalable system	Energy efficiency
Experts	-	Better climate understanding	High usability Accurate information Real time monitoring	Energy efficiency Avoid potential environmental risk

the other goal may have negative influence. Therefore, we need a process to identify and present the influence among the goals in different sustainability dimensions.

The second phase is identifying the conflict in the sustainability goal. In the climate monitoring system, performance can influence reduce energy consumption and vice versa. As mentioned in the previous section, the influence of a certain goal in different dimension can be differentiate into three type which are conflict (↓), support (↑) or neutral (→). Table 5 shows the example goal influence in the sustainability requirements. By using our proposed goal influence summary, we can make the influence visible to user in which it can help the stakeholder to make a certain decision. We can see there are some trade-off in different goals of sustainability requirements. For example, if the enterprise would like to increase the performance in term of data collection speed, it can hurt the economic sustainability in a way that it need better hardware which has higher price.

The trade-off analysis is followed by determining the priority of the sustainability requirements. The process is aimed to help the decision making process to be more efficient. Tables 6 and 7 show the applicability of the proposed method for analyzing the priority of the requirements.

The trade-off analysis result depends on the stakeholders of the company. For a company the main stakeholder can be the user. However, for another company, the main stakeholder can be the governments. Therefore, we need to identify the level of the stakeholder by assigned the weight for each stakeholder. From Table 6, we can see the company's main stakeholders is Enterprise and followed by Government and Individual user. Then, the representative of each stakeholder groups are asked to give

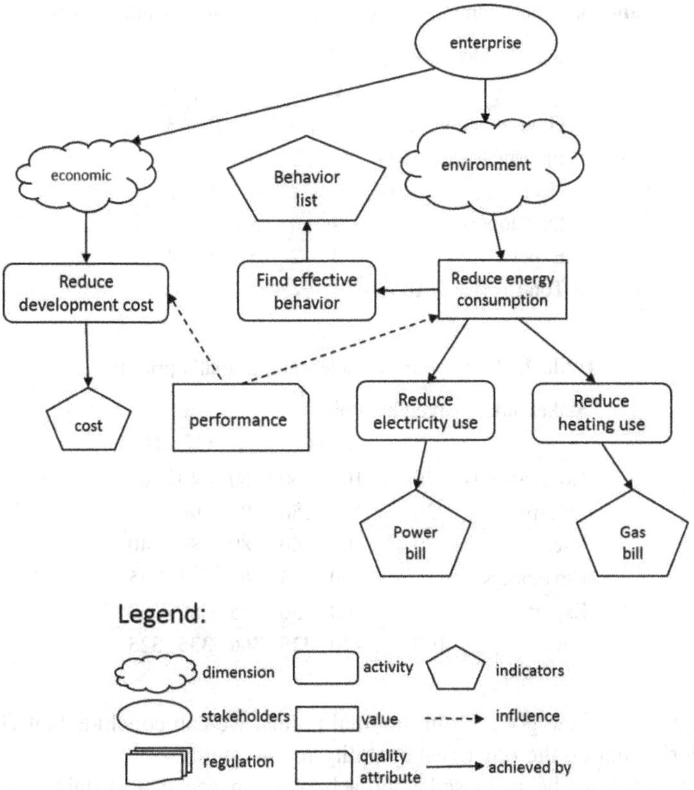

Fig. 6. Example of sustainability requirements goal model for climate monitoring

Table 5. Summary of goal influence for climate monitoring software

Goal	T	En	S	Ec
Reduce energy consumption	↑	↑	↑	↑
Reduce development cost	↓	→	→	↑
Performance	↓	↓	↑	↓
Configurability	→	→	→	↑
Find effective behavior	→	↑	↑	→

priority for the entire sustainability requirements of the system. In order to address the reliability any disagreements regarding the priority were discussed until the stakeholders group reached a consensus.

After the priority has been defined, the point is calculated by multiplying the priority and weight. Table 7 shows the point for each sustainability goal, for example the point of first goal defined by the Governments is 100 (the weight of Government is 20 and the priority of G1 given by Government is 5). The points for each goal defined by the stakeholder are summed as the total point which is used as the final consideration to

Table 6. Goals summary of goal priority of sustainability goals

Stakeholder	Weight	Priority				
		G_1	G_2	G_3	G_4	G_5
Governments	20	5	3	3	2	3
Enterprise	30	5	5	3	2	4
User	20	5	2	4	1	2
Developers	15	4	3	5	5	3
Experts	15	4	2	5	4	4
Total	100					

Table 7. Final point for determining goal's priority

Stakeholder	Weight	Point				
		G_1	G_2	G_3	G_4	G_5
Governments	20	100	60	60	40	60
Enterprise	30	150	150	90	90	120
User	20	100	40	80	80	40
Developers	15	60	45	65	65	45
Experts	15	60	30	65	60	60
Total	100	**470**	**325**	**360**	**335**	**325**

define the priority of the goal. From the total points, we can conclude that G1 has the highest priority among the other sustainability requirements.

By implementing the proposed approach, we can see that sustainability can be formally considered in the requirements engineering process. The meta model shows that even though sustainability is a high level concept, it can be decomposed as the software requirements and mapped in to the runtime model.

5 Threats of Validity

This work uses case study to evaluate the proposed approach. As mentioned in the previous section, we conducted the case study by doing some experiments with several researchers in the area of Energy Engineering. Therefore, some threats of validity cannot be avoided while conducting the experiments. This section discusses the four threats of validity which are Construct validity, Internal validity, External validity and Reliability.

Cronbach and Meehl argue that construct validity aims to investigate whether a test is to be interpreted as a measure of some attributes that are not operationally defined [16]. In this validity test, the researchers should make sure there is no reactivity bias. In order to address the construct of validity this research use the open question to reduce the bias from the respondents. In order to avoid misinterpretation of the data, we ensured that the respondent had sufficient experiments in the energy engineering. We also collected another data from different sustainability reports to mitigate the threats of construct validity.

Eisenhardt discusses that the major threat of internal validity in case study research is the confounding factor [17]. It is defined as the factor that have direct impact with the dependent and independent variables. Therefore, the testing should measure that there is an impact of the experimental condition. As we only conduct a one shot case study, there is a tendency to have the error interpreting the data and observation. In order to reduce the threats of internal validation, we conduct a one group pre-posttest design. This design is aimed to ensure the causality in the conducted experiments. In addition, we also conducted the posttest one week after the pre-test. Therefore, we could avoid the maturation as one of the threats in the internal validity.

The other threat of validity that faced in the case study is the external validity. The external should make sure that the conducted research can be generalized [18]. As mentioned in the previous section, the evaluation of the proposed approach is done using a case study design. Therefore, rather than statistical evaluation, we used a qualitative study to evaluate the proposed framework. In order to ensure the generalization, we select a group of people that have experience in requirements engineering from different domain application as the control groups.

The last threat of validity is the reliability. A study can be said as a reliable case study in which it produces same result in consistent conditions [19]. In order to mitigate the reliability issues, we conduct a formal case study design to gathered the reliable evidences based the methodology proposed by Lee and Rine [20]. The case study methodology designed research in software engineering proposed five main components which are study question, study propositions, linking data to proposition, unit of analysis and the criteria for interpreting the findings.

6 Conclusion

The proposed approach is intended to enhance the stakeholders' ability to determine the sustainability requirements of the software intensive system. The proposed framework includes a set of questions which can be used as a guideline to define the goal. By providing a meta model for the sustainability requirements of a software system, we are able to help the requirements engineer to structure the goal model of the system based on the entire sustainability dimension. In addition, we are also able to identify the potential conflict in the sustainability requirements.

One of the foreseen limitation of this proposed approach is dealing with the evaluation. The evaluation is a major issue for building a framework for sustainability. Therefore, this work relies on the case study to evaluate the proposed approach. The other limitation is the lack of focus in the social dimension part. The critical discussion of sustainability in the individual, organizational and community level is needed to get broader sustainability achievement rather than focusing only on technical and economical dimension. This work has not addressed the evaluation of the runtime model for sustainability.

The future work includes structuring a software prototype to evaluate the achievement of the sustainability requirements. The future plan will include the evaluation for the runtime model for the sustainability requirements. In addition, a further evaluation plans and generic metrics of the sustainable software engineering has been discussed and included in the future work.

Acknowledgment. This research was supported by Next-Generation Information Computing Development Program through the National Research Foundation of Korea (NRF) funded by the Ministry of Science, ICT & Future Planning (2013M3C4A7056233).

References

1. Fichter, K.: Sustainable business strategies in the Internet economy. The ecology of the new economy (2002)
2. Erdmann, L., et al.: The future impact of ICTs on environmental sustainability (2004)
3. Koomey, J.G.: Estimating total power consumption by servers in the US and the world (2007)
4. Commision, Brundtland. World commission on environment and development. Our common future (1987)
5. Tate, K.: Sustainable Software Development: An Agile Perspective. Addison-Wesley Professional, Upper Saddle River (2005)
6. Johann, T., et al.: Sustainable development, sustainable software, and sustainable software engineering: an integrated approach. In: 2011 International Symposium on Humanities, Science & Engineering Research (SHUSER). IEEE (2011)
7. Naumann, S., et al.: The greensoft model: a reference model for green and sustainable software and its engineering. Sustain. Comput. Inf. Syst. **1**(4), 294–304 (2011)
8. Dick, M., Naumann, S., Kuhn, N.: A model and selected instances of green and sustainable software. In: Berleur, J., Hercheui, M.D., Hilty, L.M. (eds.) CIP/HCC 2010. IFIP AICT, vol. 328, pp. 248–259. Springer, Heidelberg (2010). doi:10.1007/978-3-642-15479-9_24
9. Becker, C., et al.: Sustainability design and software: the karlskrona manifesto. In: Proceedings of the 37th International Conference on Software Engineering, vol. 2. IEEE Press (2015)
10. Becker, C., et al.: Requirements: the key to sustainability. IEEE Softw. **1**, 1 (2015)
11. Chitchyan, R., et al.: Sustainability design in requirements engineering: state of practice. In: Proceedings of the 38th International Conference on Software Engineering Companion. ACM (2016)
12. Mahaux, M., Heymans, P., Saval, G.: Discovering sustainability requirements: an experience report. In: Berry, D., Franch, X. (eds.) REFSQ 2011. LNCS, vol. 6606, pp. 19–33. Springer, Heidelberg (2011). doi:10.1007/978-3-642-19858-8_3
13. Penzenstadler, B., Femmer, H.: A generic model for sustainability with process-and product-specific instances. In: Proceedings of the 2013 Workshop on Green in/by Software Engineering. ACM (2013)
14. Koziolek, H.: Goal, question, metric. In: Eusgeld, I., Freiling, F.C., Reussner, R. (eds.) Dependability Metrics. LNCS, vol. 4909, pp. 39–42. Springer, Heidelberg (2008). doi:10. 1007/978-3-540-68947-8_6
15. Easterbrook, S.M.: Climate change: a grand software challenge. In: Proceedings of the FSE/SDP Workshop on Future of Software Engineering Research. ACM (2010)
16. Cronbach, L.J., Meehl, P.E.: Construct validity in psychological tests. Psychol. Bull. **52**(4), 281 (1955)

17. Eisenhardt, K.M.: Building theories from case study research. Acad. Manag. Rev. **14**(4), 532–550 (1989)
18. Berkowitz, L., Donnerstein, E.: External validity is more than skin deep: some answers to criticisms of laboratory experiments. Am. Psychol. **37**(3), 245 (1982)
19. Tellis, W.M.: Application of a case study methodology. Qual. Rep. **3**(3), 1–19 (1997)
20. Lee, S.W., Rine, D.C.: Case study methodology designed research in software engineering methodology validation. In: SEKE (2004)

MOYA: Model-Oriented Methodology for Your Awareness

Jun Hagiwara[1] and Shinobu Saito[2(✉)]

[1] NTT DATA Corporation, Tokyo, Japan
hagiwaraj@nttdata.co.jp
[2] NTT Corporation, Tokyo, Japan
saito.shinobu@lab.ntt.co.jp

Abstract. Understanding appropriate requirements is a decisive factor to project success for the enterprise system development under the complicated and changeable environments. Nowadays, the enterprise systems are expected to solve a large variety of business problems from multiple stakeholders. Therefore, requirements analysts need to understand the complicated stakeholder's business circumstances, define the issues necessary for the business purposes, and elicit requirements for the enterprise systems which could achieve the issues. In this paper, we propose a business modeling methodology which we call as MOYA that is abbreviation for "Model-Oriented methodology for Your Awareness". MOYA supports requirements analysts to elicit and organize requirements from the stakeholders. The methodology also helps stakeholders to build consensus on their own requirements. This paper describes an approach, procedures, and deliverables of MOYA. After describing an industrial project as a practice of MOYA, we discuss effectiveness and limitations of MOYA.

Keywords: Business modeling · Methodology · Stakeholder analysis · Goal analysis · Issue analysis · CATWOE · Rich picture

1 Introduction

A number of system development companies are proactively serving software frameworks, system development processes, and software-automated tools in order to strengthen their capabilities of system development. Consequently, a lot of knowledges, experiences, and implications for system development have been deeply accumulated in each company and widely shared in a system development business industry [5, 10].

However, the system development industries still have been confronted with serious problems on both increased costs and prolonged periods in their system development projects. In many cases, those problems might have been fundamentally derived from lower quality of system requirements (i.e., inaccurate and/or ambiguous requirements), which are elicited from the stakeholders of the projects [6].

In order to improve the quality of system requirements, we have developed a business modeling methodology. We call the methodology as MOYA which is abbreviation for "Model-Oriented methodology for Your Awareness". We have compiled MOYA by means of multiple requirements engineering knowledge and techniques:

S.-W. Lee and T. Nakatani (Eds.): APRES 2016, CCIS 671, pp. 68–78, 2016.
DOI: 10.1007/978-981-10-3256-1_5

Software Systems Methodology [2, 9], Goal oriented Requirements Analysis [6, 7], and Business Modeling using UML [1, 4].

NTT DATA is a global solution provider which has over 80,000 engineers in more than 40 countries worldwide. The company has owned a corporate-wide methodology for system development which is called as TERASOLUNA [8]. As a portion of the methodology, MOYA has been adopted to an early phase (i.e., requirements engineering phase) of a wide range of system development projects. In those projects, MOYA has supported the stakeholders of the projects to become clearly conscious of important issues which are necessary for their own business purposes. MOYA could also help requirements analysts to elicit and configure appropriate system requirements from the stakeholders of the projects. In this way, MOYA has been contributing to the improvement of the quality of system requirements in the system development projects at NTT DATA.

In this article, Sect. 2 describes an approach of MOYA. We also introduce the contents of analytical procedures and deliverables of MOYA. İn Sect. 3, as a practice of MOYA, we introduce the contents of an industrial system development project to which we adopted MOYA. İn the case, we developed a basic business and system plan using MOYA. In Sect. 4, we discuss the effectiveness and the limitation of MOYA based on the result of the case study. Finally, we describe conclusions and future works in Sect. 5.

2 MOYA: Model-Oriented Methodology for Your Awareness

2.1 Approach

In MOYA, we take five key elements, five techniques, and five deliverables as shown in Fig. 1. Those are defined in the MOYA. Basically, we recommend that requirements analysts focus five key elements: Stakeholder, Issue, Purpose, Business Process, and System requirements. We assume that a deeper understanding on those elements might

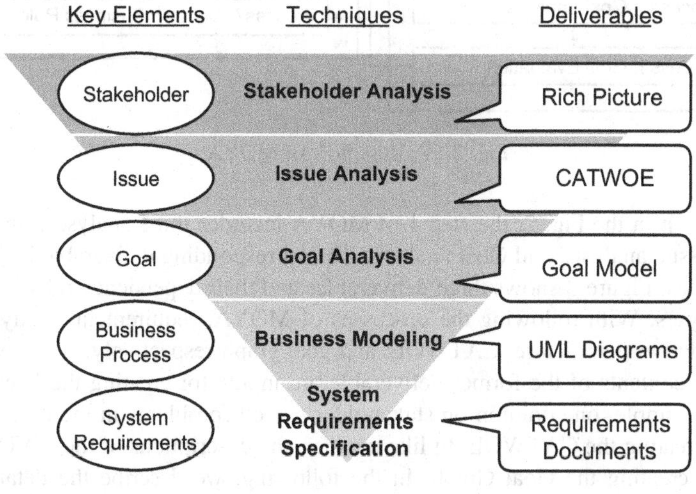

Fig. 1. Key elements, techniques, and deliverables of MOYA.

be closely linked to high-quality activities on requirements engineering phase. Consequently, those activities could contribute to the improvement of the quality of system requirements. As a result, those projects may meet the success finally.

In the approach of MOYA, requirements analysts develop the five deliverables by utilizing the five techniques in order to understand the five key elements. Figure 1 shows inter-relationships among key elements, techniques, and deliverables.

For example, as shown on the upper side of Fig. 1, requirements analysts develop the deliverable "Rich Picture" by conducting the technique "Stakeholder Analysis" to understand the key element "Stakeholder(s)". In this way, core structure of MOYA is composed of the techniques, deliverables and key elements.

2.2 Procedures and Deliverables

Figure 2 shows a process flow of MOYA. MOYA is composed of two steps: "Step 1: Define Goals" and "Step 2: Create Models". Each step includes multiple processes. Moreover, each process is composed of one or more activities.

In the step 1, requirements analysts focus to analyze three key elements: Stakeholder, Issue, and Goal. In the step 2, requirements analysts consider both business process and system requirements in order to accomplish the goals which are defined in the step 1. Needless to say, the former step is much more important rather than the later step in terms of the improvement of the quality of system requirements. In this paper, we focus to describe the contents of the step 1.

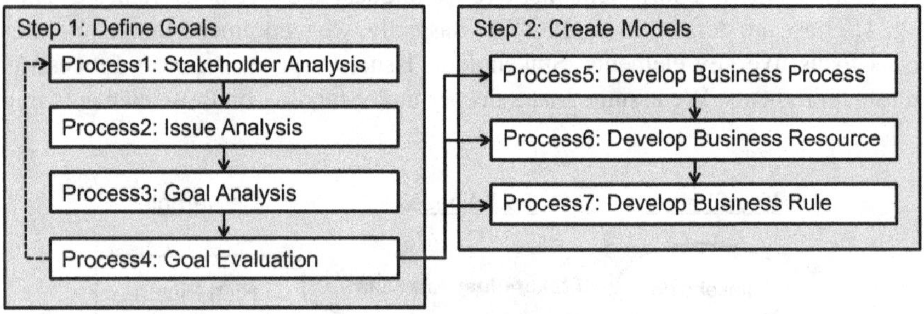

Fig. 2. Process flow of MOYA.

As shown in the Fig. 2, the step 1 of MOYA includes three analyses: Stakeholder analysis, Issue analysis, and Goal analysis. The corresponding deliverable is defined in each analysis. Figure 3 shows three deliverables and their dependency relations among three analyses. With following the processes of MOYA, requirements analysts make efforts to derive rich picture, CATWOE, and goal graph respectively. As shown in the Fig. 3, the contents of the former deliverable are inputs for creating the later deliverable. For example, one opinion on stakeholder's needs/problems in the rich picture is input for creating the CATWOE. In like manner, the descriptions of the CATWOE are inputs for creating the Goal Graph. In the following, we describe the details of the processes and the deliverables of MOYA.

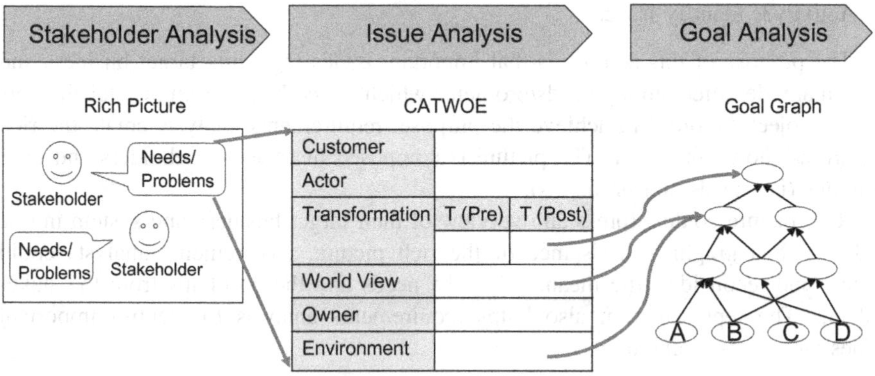

Fig. 3. Dependency relations of deliverables of MOYA.

- Process1: Stakeholder Analysis

In this process, requirements analysts try to grasp an entire situation of the target business and system. Requirements analysts elicit the stakeholders of the project, and collect needs and problems from the stakeholders.

- Activity1: Elicit Stakeholders

From the viewpoints of typical roles (e.g. system user, system operator, and system administrator) in enterprise systems, requirements analysts elicit the stakeholders from target business and system. Besides that, we recommend that top and middle managements, who are concerned with target business and system, could be important stakeholders in MOYA. The top/middle managements may have better understanding of the missions and the visions in the corporate rather than operational level stakeholders. The top/middle management's high level opinions could give insights to requirements analysts in order to catch the fundamental problems of the target business and system.

- Activity2: Collect Stakeholder's Needs and Problems

Requirements analysts conduct questionnaire survey, structured/non-structured interview, and documents analysis for collecting the needs and the problems from the stakeholders who are elicited in the previous activity.

In MOYA, there are two attention points of this activity. First point, requirements analysts should check whether the contents of the needs directly mean solution themselves or not. Stakeholders sometimes wish only solutions for themselves. However, these solutions might not be suitable to their needs. In such a situation, requirements analysts should focus to define not solutions but needs/problems. Second point, it is probable that there might be incorrect contents among opinions resulting from stakeholder's misunderstanding and/or prejudice.

To prevent from collecting incorrect contents, requirements analysts need to utilize the facts and the actual data in order to confirm whether those contents of the opinions are true or not.

– Activity3: Identify Issues

The purpose of this activity is that important issues regarding target business and system are identified among needs/problems which are collected from the stakeholders of the project. In order to achieve the purpose, requirements analysts create the rich picture as shown in Fig. 4. The picture is composed of seven stakeholders and their opinions (i.e. needs and problems).

Rich picture could express an overview of their target business and system in one slide. With a graphical assistance of the rich picture, requirements analysts could review and reconsider true meanings of the needs and the problems from the stakeholders. The representation also helps requirements analysts to identify important issues among the opinions.

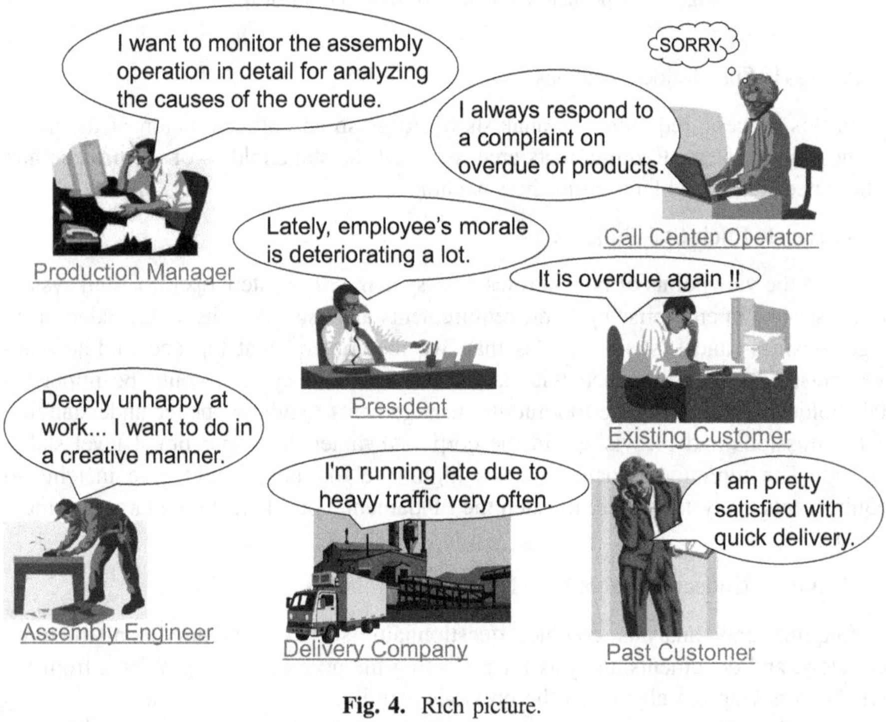

Fig. 4. Rich picture.

• Process2: Issue Analysis

Requirements analysts analyze the contents of each issue which is identified as an important in the previous process. By analyzing these issues, requirements analysts make efforts to derive the solution approaches for achieving the issues.

– Activity1: Create CATWOE

In MOYA, CATWOE [4, 5] is employed for analyzing the issues. CATWOE is abbreviation for five analytical viewpoints: Customer, Actor, pre and post

Transformation, World view, Owner, and Environment. The meanings of analytical viewpoints are following:

1. Customer: Who gives the needs from which the issue is derived?
2. Actor: Who could achieves the issue?
3. pre Transformation: What is a current situation (As-is process)?
4. post Transformation: What is a desired situation (To-be process)?
5. World view: Why do Customers regard post Transformation as desired?
6. Owner: Who could support/stop the efforts for achieving the issue?
7. Environment: What are assumptions or constraints of the efforts?

Figure 5 shows CATWOE with based on the Fig. 4 (rich picture). This issue is derived from the existing customer's opinion in the rich picture. In this way, requirements analysts could deeply understand the issues in terms of the multiple analytical viewpoints.

The contents of post Transformation, Environment, and World view are important inputs for conducting goal analysis in the next process. These inputs could be converted to the descriptions of the goals in the goal graph which be created in the process 3.

Customer	Existing Customer	
Actor	Production Manager	
Transformation	T (Pre)	T (Post)
	Taking a long time to get products ordered.	Getting products soon after placing order.
World View	Quick delivery enables to make the marketing plan on the products quickly	
Owner	President	
Environment	Assembling time can not be shorten. Delivery time depends on distance between factory and shipping address (customer office)	

Fig. 5. CATWOE.

- Process3: Goal Analysis

Based on the results of the stakeholder analysis and the issue analysis, requirements analysts conduct the goal analysis. By creating the goal graph, requirements analysts could generate the possible solutions for achieving the issues. Those possible solutions contribute to realize the post Transformation (To-be process).

- Activity1: Create Goal graph

Goal graph is tree structure which expresses relationships between upper goals and lower goals as shown in Fig. 6. Namely, accomplishment of lower goals contributes that of upper goals. In MOYA, as we mentioned before, it is recommended that the contents of the CATWOE (i.e., post Transformation, World view, and Environment) are utilized for describing the upper goals in the goal graph. Especially, the contents of post Transformation, which is regarded as the desired situation, could be converted to a description of the top goal.

As decomposing upper-goal(s) to lower-goal(s), requirements analysts elicit the lower-end goals of the goal graph. These lower-end goals could be solution options for achieving the top goals of the goal graph. For example, in the Fig. 6, the top goal "Optimizing Production Process" is finally decompose to three lower-end goals: "Shortening Production Cycle-Time", "Enhancing Machine Capacity", and "Addressing Bottleneck Process". Requirements analyst could consider that three lower-end goals are solution options.

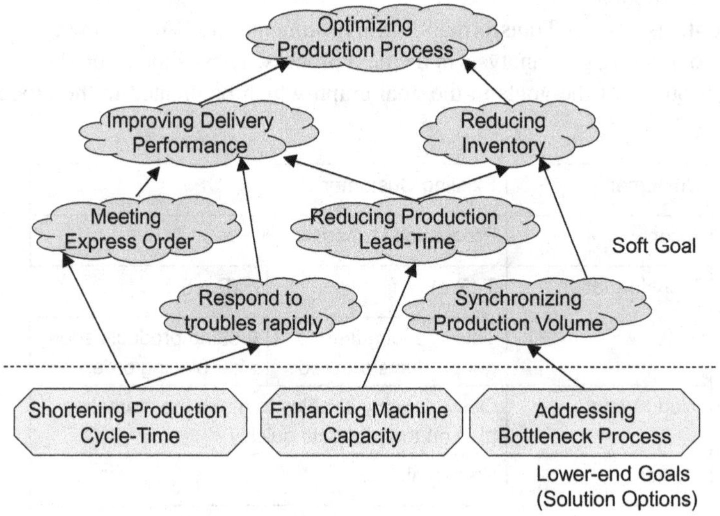

Fig. 6. Goal graph.

- Process4: Goal Evaluation

Requirements analysts define the criteria for evaluating the solution options, and finally select solution(s) for achieving the issues.

- Activity1: Define Criteria

For evaluating the solution options, MOYA suggests six fundamental criteria: Necessity, Efficiency, Effectiveness, Equalization, Priority, and Installability. By adding and modifying those criteria, requirements analysts try to define the criteria for target project.

- Activity2: Evaluate Goals

Based on the defined criteria in the previous activity, solution options are evaluated quantitatively and/or qualitatively. Requirements analysts finally select solution options for achieving the issues.

3 Case Study

We describe an industrial project as a practice of MOYA. İn this project, we have developed new ERP system for our customer who is a service provider company. As members of the requirements engineering team in the project, we have participated in the project during the basic business and system planning phase. We also have practiced MOYA in cooperation with the stakeholders.

3.1 Nature of Project

The company was established resulting from a consolidation of four former companies. Number of employees at the company is approximately 10,000. At the timing of the establishment, an ERP system was newly developed. Before the maintenance/operation contract of the existing ERP system was finished, the company launched the project for reforming the existing ERP system. The purposes of the project are "Reduction of the system operation cost", "Enhancement of the convenience of the service", and "Improvement of the efficiency of the office work". We initiated the requirements engineering team for considering and creating the basic business and system plan.

3.2 Practice of MOYA

As an initial phase of the project, we practiced MOYA (Process 1 to 4 in step 1) to create a basic business and system plan. It took approximately five months as we scheduled.

- Process1: Stakeholder Analysis

Initially, we defined four stakeholders: sales department, planning department, top-management, and information system department. Those names are corresponding to organization structure of the company.

In order to collect needs and problems of the existing ERP system and corresponding business operations, we distributed the questionnaires to over 150 employees and interviewed about 30 employees. We carefully reviewed the responses of questionnaires and the comments of the interview. After the review, we created the rich pictures. In this timing, we have re-defined six stakeholders as below.

1. Heavy system user
2. Casual system user
3. System user concerning with convenience of the system
4. System user concerning with security of the system
5. Top management
6. Information system department

- Process2: Issue Analysis

We identified 54 issues among the rich pictures which were created in the previous process. In the process, we created one deliverable (i.e., CATWOE) with correspond to one issue. Consequently, the number of CATWOE created reached into 54.

- Process3: Goal Analysis

Based on the contents from a large-set of CATWOE, we also created a large-sized goal graph which is composed of seven layer-structures. The head layer of the goal graph consists of three top-goals. For example, one of three top-goals is "Ensure system security and Protect system user from security threats". The top goal is decomposed five lower-end (second layer) goals. The descriptions of the goals are following.

1. "Distribute anti-virus software to all machines and servers"
2. "Upgrade the security patch software automatically"
3. "Monitor the security status of all user client-machine automatically"
4. "Shorten the expiration period of user authentic password"
5. "Introduce Biometrics authentication"

- Process4: Goal Evaluation

As evaluation criteria, we employed the six fundamental criteria which are suggested in MOYA. We selected the qualitative approach for goal evaluation because it was difficult for us to set the values of thresholds. Each lower-level goals (i.e., solution options) are evaluated on three grades: A (good), B (fair), and C (poor).

Moreover, we selected the goals regardless of result of three grades evaluation because lower-level goals possessed specific traits. Specific traits are described as below.

1. Have a synergistic effect with other solution options
2. Have a numerous impact on competitor, market, and customer

- Result of MOYA

After the practice of MOYA, top management of the company finally authorized the solution options (lower-level goals that were selected by the goal evaluation).

3.3 Feedbacks from Stakeholders

After the completion of practice of MOYA, we collected feedbacks from the stakeholders who have participated in the activities of MOYA. Summaries of the feedbacks are followings. We categorize the feedbacks in terms of two aspects: Positive and Negative.

- Positive Feedback
  - As creating the goal graph, we could visualize the structure of a number of goals. The image helps us to grasp the corresponding-relations between the end (upper goals) and the means (lower goals).
  - With reference to the goal graph and their goal evaluation by the criteria, we could select and configure the solution options (lower-end goals) with aim to promote the achievement of total optimization for new business and system.
  - The goal model could clarify and record the deliberation processes for creating the solution options (lower-end goals) from the desired situation (top goals).

- Negative Feedback
  - It took too much time to analyze a massive amount of the stakeholder's opinion.
  - The complexity of the CATWOE and the goal graph made us have hardly feeling of the advantage of the deliverables.

4 Discussions

4.1 Effectiveness of MOYA

With based on the result of the case study and the feedbacks from the stakeholders, we claim that the visualization-effects from the deliverables defined in MOYA could contribute to make the stakeholders have shared clear awareness of important issues. MOYA also supports to review the solution options. With reference to the goal graph and the CATWOE, stakeholders could reconsider whether these solutions options (i.e., lower-end goals) contribute the upper goals or not. Moreover, after selecting the solution options, the rich picture allowed the requirements analysts to validate the result of the selections. They could carefully confirm whether the selected solution options could solve the problems or not.

4.2 Limitations of MOYA

Some stakeholders have referred the negative feedbacks on MOYA. We assumed that the small number of the participants in the activities of MOYA is the reason why those stakeholders had negative feelings on MOYA. As described in the case study, we have analyzed approximately 180 opinions (about 150 questionnaires and 30 interviews) from the employees in the company. On the one hand, only a few employees were assigned to our requirements engineering team in the project. The majority of the employees did not participate in the activities of MOYA. They, who are non-participation, only look at the deliverables after the completion of the activities of MOYA. We have to acknowledge that the practice of MOYA is time-consuming and high-complexity. Without the derivation process of the deliverables, they hardly understand the deliverables. They did not feel the values of the activities of MOYA.

4.3 Capability for Practice of MOYA

We have realized that the soft-skills are important for requirements analysts who practice MOYA. In order to involve and motivate the stakeholders of the project with the activities in MOYA, requirements analysts need to acquire communication technique and knowledge. We have developed an instruction course which is a two-day workshop style. As introducing MOYA and its training course, we are going to cultivate the requirements engineer's capabilities for practice of MOYA.

5 Summary

In this paper, we have proposed MOYA that our business modeling methodology. MOYA supports stakeholders to become conscious of issues necessary for their own purposes. MOYA also helps requirements analysts to elicit and understand appropriate system requirements from the stakeholders, and improve the quality of system requirements. As a practice of MOYA, we have conducted the basic business and system planning phase on the basis of the procedures and deliverables of MOYA. From the result of the case study, we discussed the effectiveness and the limitations of MOYA.

References

1. van Lamsweerde, A.: Requirements Engineering: From System Goals to UML Models to Software Specifications. Wiley, New York (2009)
2. Willson, B.: Soft Systems Methodology: Conceptual Model Building and Its Contribution. Wiley, New York (2001)
3. Eliksson, H., Penker, M.: Business Modeling With UML: Business Patterns at Work. Wiley, New York (2000)
4. i star framework. http://www.cs.toronto.edu/km/istar/
5. IBM Rational. http://www-01.ibm.com/software/jp/rational/
6. JISA REBOK WG (ed.): Requirements Engineering Body Of Knowledge (REBOK), Ver. 1.0, Kindaikagakusha (2011). (in Japanese)
7. Fowler, M.: UML Distilled: A Brief Guide to the Standard Object Modeling Language. Addison Wesley Professional, Reading (2003)
8. NTT DATA. http://www.terasoluna.jp/
9. Checkland, P., Scholes, J.: Soft Systems Methodology in Action. Wiley, New York (1999)
10. SDAS. http://www.fujitsu.com/jp/solutions/infrastructure/dynamic-infrastructure/sdas/

2-dip-dfs: Algorithm to Detect Conflict Between Two Goal Selection Criteria

Shin'ichi Sato(✉)

Department of Industrial and Systems Engineering, College of Science and Engineering, Aoyama Gakuin University, Tokyo 252-5258, Japan
sato@ise.aoyama.ac.jp

Abstract. In the requirements elicitation stage of goal-oriented requirements engineering, there is often a conflict between two goal selection criteria. This conflict can be detected by analyzing all pairs of each goal selection criterion for each subgoal graph with two layers in a goal graph. In this study, we defined a conflict between two goal selection criteria as a situation in which the selected child goal of a parent goal is different between one goal selection criterion in the pair and the other. On this basis, we proposed a metric called *2-dip*(g) for detecting conflicts. Based on *2-dip*(g), we developed an algorithm that can detect all conflicts in a goal graph. Because the algorithm uses a depth-first search, we called it *2-dip-dfs*. By applying the algorithm result to a real goal graph, we demonstrated that *2-dip-dfs* functions as designed.

Keywords: Goal-oriented requirements engineering · Requirements elicitation · Goal selection criteria · Conflict detection

1 Introduction

Goal-Oriented Requirements Engineering (GORE) has received considerable attention as a requirements elicitation method in Requirements Engineering (RE) research. As the expectations of GORE have been increased, numerous goal models have been proposed [15] and various analysis methods of these models have been developed [5]. Among these analysis methods, the current study focuses on algorithms. Existing studies of formal goal selection in GORE research fundamentally assume that a goal graph (see Sect. 2) is a huge propositional logic formula that can be denoted as a *conjunctive normal form (CNF)*[1] [16]. For a propositional logic formula forming CNF, the problem of asking whether the variables of a given Boolean formula can be replaced consistently by values of true or false is called a propositional/Boolean satisfiability/testing (SAT) problem.

A range of SAT-based goal selection algorithms have been studied intensively in RE research [16]. These are called *goal reasoning* algorithm, and are based on the

[1] A Boolean fomula is in CNF if and only if it is in the form $\bigwedge_i \bigvee_j l_{ij}$, where l_{ij} are *literals*. A literal is an atomic formula or its negation. A disjunction $\bigvee_j l_j$ is called *clause*.

S.-W. Lee and T. Nakatani (Eds.): APRES 2016, CCIS 671, pp. 79–93, 2016.
DOI: 10.1007/978-981-10-3256-1_6

fact that a goal can be considered as a proposition that has the value of being true or untrue. Existing goal reasoning algorithms focus only on the logical binary operators *and* and *or*. However, this approach cannot concretely identify goals that are finally achieved if they are in an OR relation, and the ensuing ambiguity can cause disagreements among stakeholders with conflicting requirements. This has a negative effect on consensus building, which serves to increase costs in the requirement analysis phase. The cost of fixing software defects exponentially increases if the development phase becomes part of the next software life cycle [3]. For this reason, the selection of goals for an OR relation in the requirements elicitation phase of GORE approach is important for avoiding the so-called "death-march project." Latent conflict detection methods for GORE need to be developed, and these should be automated as far as possible to reduce the cost of manual effort and avoid errors introduced by these manual efforts.

We envisaged the goal graph structure as a directed acyclic graph (DAG). In GORE research, various attributes have been proposed for different applications [6,7,14,16]. In this study, each above mentioned attribute was treated as a goal selection criterion. To achieve this, goal selection can be performed by taking the value of each goal selection criterion without referring to the truth value of each goal. This enables goal selection in an OR relation by prioritizing goals based on the values of the goal selection criteria (a detailed treatment is reported in Sects. 2.4 and 2.5). However, if multiple goal selection criteria are considered, a conflict can arise among selected goals. This conflict must therefore be identified in goal selection. Based on this understanding, we proposed an algorithm that can detect conflicts between two goal selection criteria.

The paper is organized as follows. Section 2 provides an overview of the GORE method, reviews the existing goal selection criterion *Cov*(g), and proposes a formal metric for the definition of goal selection criterion *Vcn*(g). On this basis, a conflict detection metric for a goal g is defined, which we call *2-dip*(g). Section 3 proposes a conflict detection algorithm *2-dip-dfs* using *Cov*(g) and *Vcn*(g). Section 4 presents an experiment done to evaluate *2-dip-dfs* and discusses the results. Section 5 analyzes the difference between existing conflict detection methods for use with GORE and our proposed method. Finally, in Sect. 6, conclusions are reported and suggestions are made for future research.

2 Overview of GORE

GORE is an established method of requirements elicitation [14,16]. The basic procedure of GORE is as follows.

1. Based on customers' needs, *initial goals* are formulated. A *goal* is an achieved state and is usually represented as an oval in natural language.
2. Goals are recursively decomposed until concrete goals are defined that can be processed by a software-intensive information system developed by stakeholders.
3. Goals are selected as requirements from well-decomposed *final goals* on the goal graph.

Through this procedure, a graph is developed with goals as elements. This is called a *goal graph* and is the final artifact of GORE. The *edge (arc)* of the goal graph shows a decomposition (refinement) relation. Between two goals that are mutually adjacent through an edge, the pre-decomposed one is called the *parent goal* and the other one is called the *child goal*. It is possible to produce multiple child goals from a single parent goal. The decomposition relation may be either *AND decomposition* or *OR decomposition*. In AND decomposition, if all child goals are achieved, then their common parent goal is achieved. It is therefore said that an *AND relation* exists among child goals made by AND decomposition. In OR decomposition, if at least one child goal is achieved, then the parent goal is achieved. It is therefore said that an *OR relation* exists among child goals made by OR decomposition. Each edge is represented by an arrow drawn from a parent goal to child goal. In other words, the root of an edge is a parent goal and the head is a child goal. From the viewpoint of graph theory, an edge can be regarded as an *ordered pair*. By definition, the ordered pair of a parent goal p and child goal c, with the first coordinate p and second coordinate c, is a two tuple denoted by (p, c). A DAG with AND decomposition and OR decomposition is called an *AND/OR graph* [11]. A goal graph is an AND/OR graph.

Figure 1 [9] shows an example of a goal graph or more concretely, a premature goal graph on the route to a complete goal graph. In Fig. 1, ovals (nodes) and arrows (directed edges) express goals and mutual decomposition relations respectively. An arc between edges shows that the edges are mutually decomposed by AND decomposition. In contrast, if no arc is attached between edges, then they are decomposed by OR decomposition. In Fig. 1, for example, "Schedule meeting" is achieved only if both its child goals ("Collect timetables" and "Choose schedule") are achieved because these child goals are adjacent to the common parent goal "Schedule meeting" through AND decomposition. In contrast, "Choose schedule" is achieved if either "Manually" or "Automatically" is achieved because these child goals are adjacent to the common parent goal "Choose schedule" through an OR decomposition.

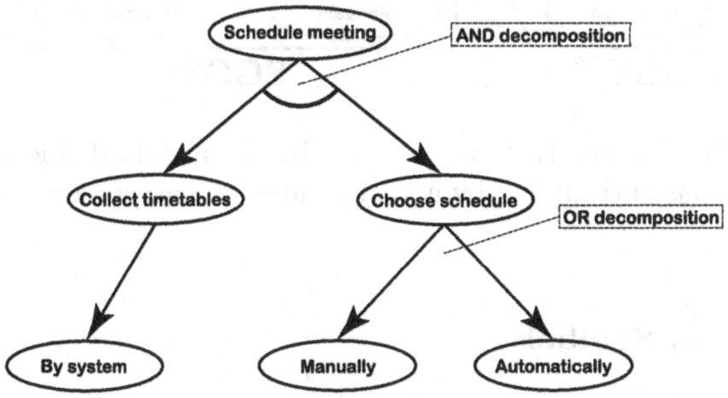

Fig. 1. Example of a partial goal graph [9].

2.1 Definitions of Terms

1. *Initial goal (Root goal)*: A goal that is made first in GORE based on customers' needs. It is possible to set multiple initial goals.
2. *Final goal (Leaf goal)*: A goal that has no child goal at all. A final goal is a goal that is sufficiently concrete to be achieved by a software-intensive information system developed by stakeholders.
3. *Sub goal graph (SGG)*: A part of a goal graph.
4. *Sub goal graph with two layers (SGG_2)*: An SGG consists of one parent goal and its child goals. An SGG_2 is clearly a minimal constitutional unit of any SGG. SGG_2s are divided into the following three types:
   (a) SGG_2^{AND}: An SGG_2 in which the parent goal is adjacent to its child goals by AND decomposition.
   (b) SGG_2^{OR}: An SGG_2 in which the parent goal is adjacent to its child goals by OR decomposition. If an SGG_2 has only one child goal, then it is regarded as an SGG_2^{OR}.
   (c) $SGG_2^{AND/OR}$: Each of the SGG_2s, but excluding both $\mathrm{SGG}_2^{\mathrm{AND}}$ and $\mathrm{SGG}_2^{\mathrm{OR}}$.

Figure 2 presents each constitutional unit of the goal graph from Fig. 1. Figure 1 shows one $\mathrm{SGG}_2^{\mathrm{AND}}$ and two $\mathrm{SGG}_2^{\mathrm{OR}}$s. No $\mathrm{SGG}_2^{\mathrm{AND/OR}}$ is present in Fig. 1.

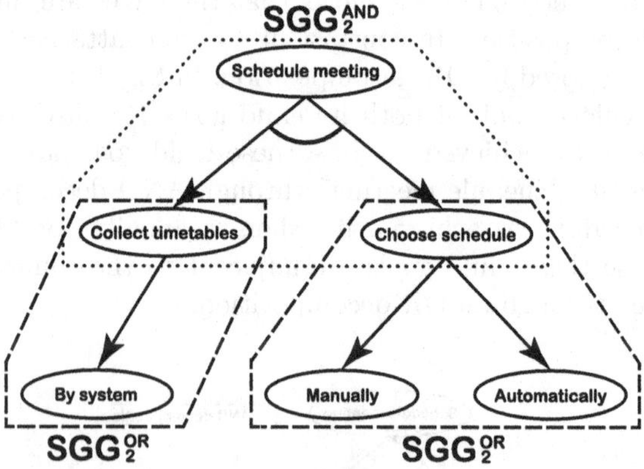

Fig. 2. Three SGG_2s in Fig. 1. SGG_2 surrounded by a dotted line is categorized as $\mathrm{SGG}_2^{\mathrm{AND}}$. Two SGG_2s surrounded by a dashed line are categorized as $\mathrm{SGG}_2^{\mathrm{OR}}$.

2.2 Definition of Symbols

Presuming that G denotes a set of all goals in a goal graph and that E denotes a subset of the Cartesian product between G and G itself ($E \subset G \times G$), then a goal graph without decomposition relations is defined by the two tuple (G, E). Elements of G and E are denoted by g and e ($g \in G$, $e \in E$), respectively.

2.3 Contribution Value

Several metrics are used as goal selection criteria. From existing metrics, this study adopts a *contribution value* [6] as the goal selection criterion because many GORE methods have used similar metrics. A contribution value is attached to an edge and its value is defined as an integer from −10 to +10. The value indicates the contribution that a child goal makes to the achievement of its parent goal: the larger the value, the greater the contribution. A negative value means that the child goal blocks the achievement of its parent goal. Using the definition of AND decomposition and OR decomposition from Sect. 2, different values can be set to each edge in an OR relation, but the same value must be set to all edges in an AND relation. In this study, the contribution value of an edge e is denoted by $Cov(e)$, while e is defined by an ordered pair (g, g') (a contribution value from a goal g' to goal g). By the definition of a goal graph, a child goal may have multiple parent goals. However, if there is only one parent goal for a given child goal, then e is equal to (g, g'). This is a proof that a relational expression between e and (g, g') is given by the following formula:

$$\exists ! e \in E \text{ s.t. } e = (g, g') \Rightarrow Cov(g') := Cov(e) \tag{1}$$

Figure 3 presents an SGG (see Sect. 2.1) in which a contribution value is attached to each edge. Each contribution value in this goal graph is arbitrarily set by the author. For an analysis of the proposed algorithm, in this figure, *indices* are attached to each goal, as well as contribution values.

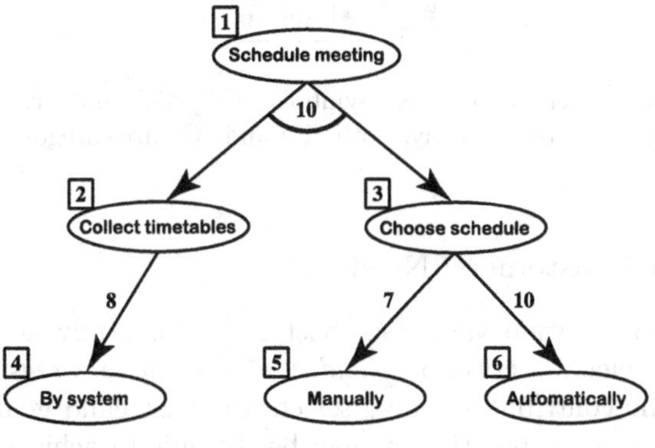

Fig. 3. Figure 2 with contribution value. Contribution values and indices are set to edges and goals, respectively.

2.4 Preference Matrix

It is important to consider the extent to which each stakeholder is satisfied by the achievement of a goal. As an attribute to represent this degree of satisfaction, a *preference value* has been proposed [6,7]. Similar to the contribution value, the preference value is defined as an integer from −10 to +10 (see Sect. 2.3). The value reflects the degree of satisfaction for a stakeholder when a child goal is achieved. To measure the degree of coincidence of preference values among stakeholders, the preference value should be derived from the viewpoints of all stakeholders. For this reason, preference values are usually attached to a particular goal in the form of a *preference matrix* [6,7].

Figure 4 shows an example of a preference matrix. Robertson et al. [12] identified seven stakeholders in the requirement process: "client," "user," "customer," "operator," "administrator," "owner," and "engineer." In this figure, three main stakeholders are selected, "user," "owner," and "engineer" because consensus building among these three stakeholders is a key factor in the success of a project. Note that both "user" and "owner" are regarded as a customers in this study. The degree of mutual influence will normally differ among these stakeholders [2]. To ensure that differences in influence do not reflect the preference value, no stakeholder can see the value set by any other stakeholder. In Fig. 4, for example, "user" cannot see components in either the second row, $(10, 10, -10)$, set by "owner" or the third row, $(5, -10, 0)$, set by "engineer." Only *requirement analysts*, the third stakeholder who is independent of other stakeholders, can examine and analyze the value of all components.

estimator \ the estimated	U	O	E
U	8	-7	0
O	10	10	-10
E	5	-10	0

Fig. 4. Example of preference matrix. Symbols "U," "O," and "E" indicate "user," "owner," and "Engineer" respectively. Both "U" and "O" are categorized as customers.

2.5 Validity of Customers' Needs

Alongside the contribution value (see Sect. 2.3), this study adopted a second metric as a goal selection criterion: *validity of customers' needs.* Goal selection based only on the contribution value selects the best child goal to achieve its parent goal. However in practice, it may be difficult to achieve the child goal because of some unique constraints in the organization to which stakeholders belong. Conversely, it may be the case that although a child goal is difficult to achieve under present circumstances and its contribution value is low, the achievement of child goal is essential from the viewpoint of management strategy. For these reasons, goal selection should consider not only the contribution value but also the validity of achieving the child goal.

GORE is a method for eliciting the requirements of the customers of a subset of stakeholders. When using GORE, a goal selection method that considers the validity of customers' needs is required. In this study, the *Validity of customers' needs* of a goal g is defined as follows:

$$Vcn(g) := \frac{\sum_{\substack{s \in S \\ c \in C}} p(g)_{s,c}}{|S| \times |C|} \tag{2}$$

where S denotes a set in which each element is a different stakeholder and C denotes a set of stakeholders that fall under customers; thus, $C \subset S$. As described above, G denotes a set of all goals in a goal graph (see Sect. 2.2); thus, $g \in G$ denotes a goal. $p(g)_{s,c}$ denotes the preference value (see Sect. 2.4) of a customer $c \in C$ from $s \in S$'s view of a goal g. More concretely, the preference value represents an estimate value that a stakeholder sets from the viewpoint of a customer under the assumption of belonging to the customer side. For example, Fig. 5 represents an SGG_2^{OR} (see Sect. 2.1) set using both a contribution value (Sect. 2.3) and preference matrix (Sect. 2.4) as goal selection criteria. For analysis, *indices* are also attached to each goal. If the validity of customers' needs is set as a goal selection criterion, "hiring a local interpreter" is selected because $Vcn(3) = -0.5$ is larger than $Vcn(2) = -7.5$, which has the largest value among child goals. The value of $Vcn(g)$ is defined as an integer from -10 to $+10$ in the same manner as that of $Cov(g)$. This means that the values of $Cov(g)$ and $Vcn(g)$ can be quantitatively compared. In goal selection for an SGG_2^{OR}, the child goal that has the largest value for both $Cov(g)$ and $Vcn(g)$ should be selected. However, a case may arise in which one is larger than the other. We call this conflict situation *2-dip*, an abbreviation of "differences in priority between two goal selection criteria." Figure 5 shows an imaginary example of SGG_2^{OR} with 2-dip, and Table 1 lists the values of $Cov(g)$ and $Vcn(g)$. From Table 1, it can be confirmed that the selected child goal differs between $Cov(g)$ and $Vcn(g)$ and that 2-dip is present in the SGG_2^{OR} of Fig. 5.

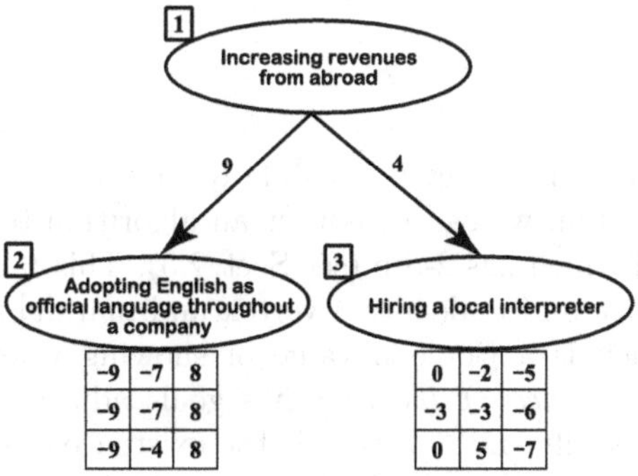

Fig. 5. Example of SGG_2^{OR} with contribution values and preference matrices.

Table 1. Values of each *Cov*(*g*), *Vcn*(*g*), and *2-dip*(*g*) in Fig. 5. *Cov*(*g*) is a contribution value of goal *g* to its a parent goal. *Vcn*(*g*) is a validity of customers' needs of goal *g*. *2-dip*(*g*) is a conflict detection metric reflecting the truth value that at least one 2-dip exists for goal *g*.

Goal	$Cov(g)$	$Vcn(g)$	$2\text{-}dip(g)$
2	9	−7.5	⊤
3	4	−0.5	

2.6 Metric to Detect Conflict

In this study, we identified the existence of 2-dip (Sect. 2.5) from the following logical statement:

$$2\text{-}dip(g) = \top :\Leftrightarrow 2\text{-}dip(R) \neq \varnothing \tag{3}$$

where *2-dip*(*g*) is a function from G to $\{\top, \bot\}$ and *2-dip*(*R*) is a *graph* defined as follows:

$$2\text{-}dip(R) := \{ (c, c') \in Sub(g) \times Sub(g') \mid cRc' \} \tag{4}$$

In this formula, c is a child goal of g, and $Sub(g) := \{ g' \mid (g, g') \}$ is a set that includes all child goals of g as elements. cRc' is a *binary relation* over $Sub(g)$ defined as follows:

$$\begin{aligned} cRc' :\Leftrightarrow \Big(&\big(Crt(c) < Crt(c')\big) \wedge \big(Crt'(c) > Crt'(c')\big) \\ &\vee \big(Crt(c) > Crt(c')\big) \wedge \big(Crt'(c) < Crt'(c')\big)\Big) \wedge c \neq c' \end{aligned} \tag{5}$$

In this formula, $Crt(c)$ denotes any function that gives a value of a goal selection criterion for goal c such as $Cov(c)$ (see Eq. (1)) or $Vcn(c)$ (see Eq. (2)).

In summary, if a goal g has at least one pair of child goals of g and 2-dip exists in the pair, then *2-dip*(*g*) is true. Otherwise, it is false.

3 *2-dip-dfs*

Depth-first search (dfs) [1] makes it possible to visit all goals in a goal graph. Based on a dfs algorithm, we can construct an algorithm to detect all goals in a goal graph, each of which has 2-dip (see Sect. 2.5). This *2-dip-dfs* algorithm is shown in Fig. 6, both v and w denote a vertex and mark[v] is a list (an array), each element of which is a Boolean value of showing whether v is *unvisited* or *visited*. $L[v]$ is an *adjacency list* that has goals adjacent to v as elements. The difference between dfs and *2-dip-dfs* is the existence of the procedure *2-dip* (Fig. 6, line 6). Apart from that, *2-dip-dfs* is exactly the same as dfs.

Figure 7 shows the detail of the procedure *2-dip*. The conditional expression of statement 11 ($n\_and = 0$ **and** $n\_or >= 2$) judges whether v connects its child goals by OR decomposition. If this expression is true, then v is the target goal

```
1: procedure 2-dip-dfs ( v: vertex );
2: var
3:    w: vertex;
4: begin
5:    mark[v] := visited;
6:    2-dip(v);
7:    for each vertex w on L[v] do
8:       if mark[w] = unvisited then
9:          2-dip-dfs(w)
10: end; { 2-dip-dfs }
```

Fig. 6. *2-dip-dfs.*

```
procedure 2-dip ( v: vertex );
var
   w: vertex;
   n_and, n_or: integer; n_and := 0; n_or := 0;
begin
   for each vertex w on L[v] do
      if w is a child goal of v and both w and v are in an AND decomposition then
         n_and := n_and + 1
      else if w is a child goal v and both w and v are in an OR decomposition then
         n_or := n_or + 1;
   if n_and = 0 and n_or >= 2 then begin
      { v is the target of detection (is a parent goal in SGG_2^OR). };
      if has2-dip(v) = 1 then
         { v is a goal which has at least one 2-dip. }
      else
         { v is a goal which has not 2-dip at all. }
   else
      { v is a goal which is not needed to detect. }
   end
end; { 2-dip }
```

Fig. 7. *2-dip* (procedure of statement 6 in Fig. 6).

for which 2-dip is detected. A function *has2-dip* is then called to test for the existence of 2-dip. *has2-dip* is designed to examine all pairs of each child goal, except for the pair of a child goal and itself in an $\mathrm{SGG}_2^{\mathrm{OR}}$. If there is at least one difference in priority between two goal selection criteria, *has2-dip* returns the value 1; otherwise, it returns 0. If 1 is returned, then v has 2-dip; if 1 is not returned, then v does not have 2-dip. The detail of *has2-dip* is shown in Fig. 8. The contents of *has2-dip* differ according to the two goal selection criteria set. Figure 8 shows a particular case of *has2-dip*, $\textit{has2-dip}_{cov,\ vcn}$ in which both the contribution value (Sect. 2.3) and validity of customers' needs (Sect. 2.5) are presupposed to be set as goal selection criteria.

In Fig. 8, a number larger than the maximum number of child goals (the total number of goals excluding initial goals) in a goal graph is set to M in advance, $cov(w)$ is a function that returns the contribution value to w of v, and $vcn(w)$ is a function that returns the value of validity of customers' needs of w. $idx(w)$ is a function that returns the index number of w. Note that each index of w is the

```
function has2-dip ( v: vertex ): integer;
var
   w: vertex;
   k, l, m, con: integer; k := 0; con := 0;
   L_cv: array [0..M + 1] of integer;
   L_vc: array [0..M + 1] of real;
   L_id: array [0..M + 1] of integer;
begin
   for each vertex w on L[v] do
      if w is a child goal of v then begin
         k := k + 1;
         L_cv[k] := cov(w);
         L_vc[k] := vcn(w);
         L_id[k] := idx(w)
      end;
   for l := 1 to k do begin
      for m := 1 to k do begin
         if l < m then
            if ((L_cv[l] > L_cv[m]) and (L_vc[l] < L_vc[m])) or ((L_cv[l] < L_cv[m])
and (L_vc[l] > L_vc[m])) then begin
               { Difference in priority between L_id[l] and L_id[m] exists. };
               if con = 0 then
                  con := 1
            end
            else
               { No difference in priority between L_id[l] and L_id[m] exists. };
         m := m + 1
      end;
      l := l + 1
   end;
   return con
end; { has2-dip }
```

Fig. 8. *has2-dip*$_{cov,\ vcn}$. Note that this function is for a particular case of *has2-dip* (statement 13 in Fig. 7) as both the contribution value and the validity of customers' needs were presupposed to be set as the goal selection criteria.

element of the set of all natural numbers when 0 is excluded. Each of *L_cv*, *L_vc*, and *L_id* is a list, the elements of which are *cov*(*w*), *vcn*(*w*), and *idx*(*w*) respectively.

4 Evaluation Experiment

A computer program written in C was used to evaluate *2-dip-dfs* (Fig. 6). The code comprised approximately 200 lines, excluding comments. Note that the size of an adjacency list representing a goal graph depends on the size of the graph. In the program, each contribution value *Cov*(*g*) (see Eq. (1)) and validity of customers' needs *Vcn*(*g*) (see Eq. (2)) were set as a function giving the value of the goal selection criterion. To confirm the practicality of *2-dip-dfs*, we applied it to the small goal graph shown in Fig. 9, which was based on a real project. This goal graph was designed for the reduction of production costs in Company A [13]. Each value of the validity of customers' needs was arbitrarily set by the author. Figure 10 shows the results. Six SGG_2^{OR}s (see Sect. 2.1) of each parent goal needed to be detected as the target. Table 2 compares both the contribution value and the validity of customers' needs in each SGG_2^{OR} as well as the result of 2-dip detection. A comparison between Fig. 10 and Table 2 confirms that *2-dip-dfs* appropriately functioned for all SGG_2^{OR}s in this experiment, as shown in Fig. 9.

Although *2-dip-dfs* was designed to treat only SGG_2^{OR} in the category of SGG_2 (see Sect. 3), this is insufficient for 2-dip detection, and *2-dip-dfs* should be extended to treat $SGG_2^{AND/OR}$ (see Sect. 2.1) as well as SGG_2^{OR}. In a $SGG_2^{AND/OR}$, a pair of child goals is in an AND relation with each other, and any other child goals are regarded as being in an OR relation. Goal selection is necessary in such cases. However, this is outside the scope of the current study, and *2-dip-dfs* was not provided with functions to treat such cases. A comparison between Figs. 9 and 10 shows an $SGG_2^{AND/OR}$ (parent goal is 4) that *2-dip-dfs* failed to detect as a target.

5 Discussion

Managing conflicts is an important goal of GORE [14,16]. Several studies [4,8] have proposed formal conflict detection methods for KAOS [14] and Tropos [10]. However, as these methods have been developed for specific GORE models, their general applicability has been unclear. In addition, these methods were developed to analyze conflicts among goals and not among goal selection criteria. These approaches have therefore focused on the contents of each goal and on its truth value. In contrast, *2-dip-dfs* (see Fig. 6) is interested in the goal selection criterion and focuses on attaching a quantitative value to it. This is because the value of the goal selection criterion represents the worth of the goal from a particular viewpoint. Although *2-dip-dfs* and studies such as [4,8] have the aspect of conflict detection in common, they target different aspect of a conflict.

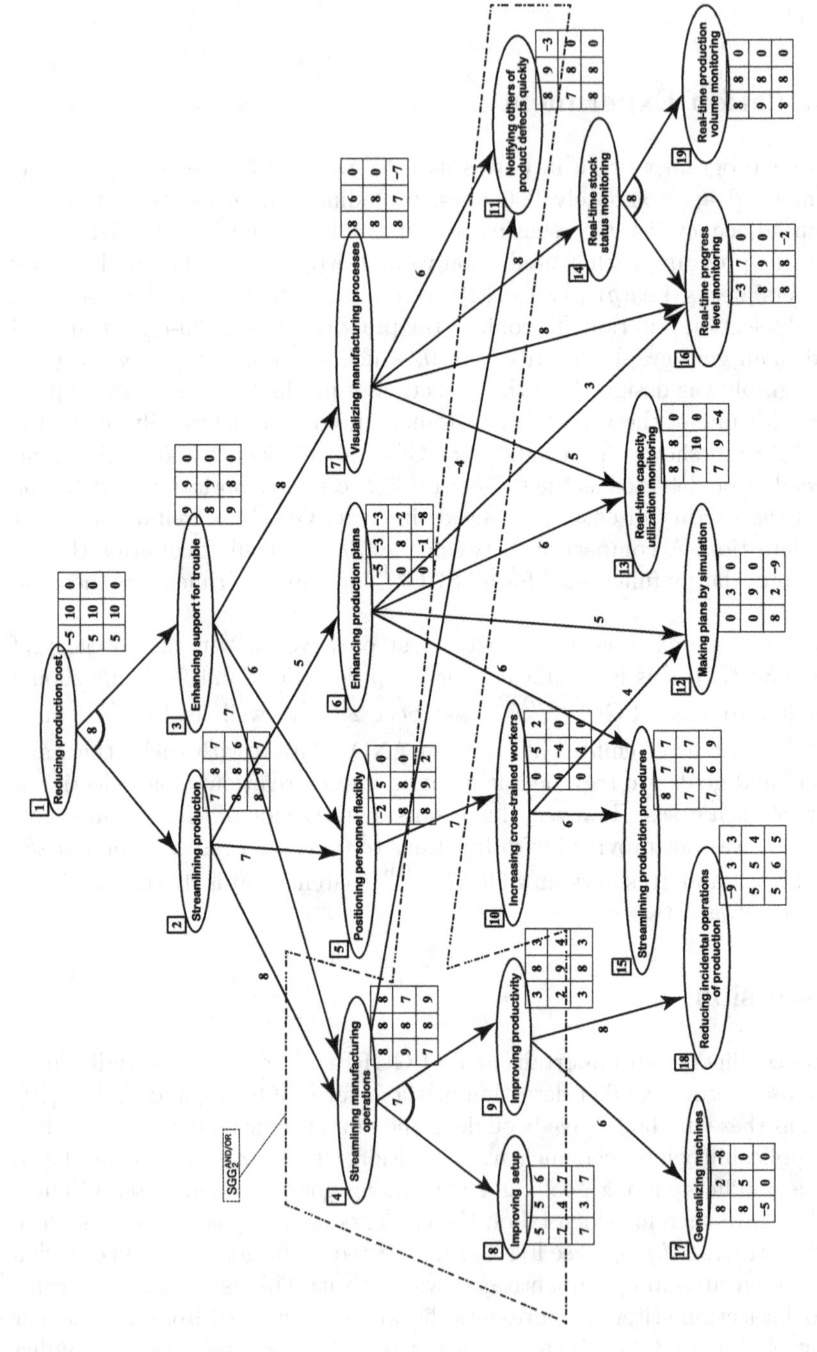

Fig. 9. Real goal graph [13].

```
2␣//␣target
3␣//␣target
␣␣␣␣␣␣␣␣(4,␣7)␣//␣dip␣exists
␣␣␣␣␣␣␣␣␣␣␣␣␣␣␣␣3␣//␣2-dip
10␣//␣target
6␣//␣target
␣␣␣␣␣␣␣␣(12,␣16)␣//␣dip␣exists
␣␣␣␣␣␣␣␣␣␣␣␣␣␣␣␣6␣//␣2-dip
7␣//␣target
␣␣␣␣␣␣␣␣(11,␣13)␣//␣dip␣exists
␣␣␣␣␣␣␣␣(11,␣14)␣//␣dip␣exists
␣␣␣␣␣␣␣␣(11,␣16)␣//␣dip␣exists
␣␣␣␣␣␣␣␣(13,␣14)␣//␣dip␣exists
␣␣␣␣␣␣␣␣(13,␣16)␣//␣dip␣exists
␣␣␣␣␣␣␣␣␣␣␣␣␣␣␣␣7␣//␣2-dip
9␣//␣target
␣␣␣␣␣␣␣␣(17,␣18)␣//␣dip␣exists
␣␣␣␣␣␣␣␣␣␣␣␣␣␣␣␣9␣//␣2-dip
```

Fig. 10. Application results for the goal graph of Fig. 9 of *2-dip-dfs*. Note that each comment in *2-dip* (see Fig. 7) and *has2-dip* (see Fig. 8) is replaced by a simplified one. In the left column, parent goals in a certain $\mathrm{SGG}_2^{\mathrm{OR}}$ are printed for the target goals of conflict detection. In the center column, if there is a difference in priority between two goal selection criteria for two child goals of a target goal, the pair of two child goals is printed. In the right column, goals that have 2-dip are printed.

By focusing on each goal selection criterion, conflict detection can be viewed as identifying the difference in priority attached to goal selection criteria corresponding to each goal. This difference in approach enables conflict detection to be simplified. To identify the truth value of a certain goal, natural language processing (NLP) methods must be applied to enable a semantic analysis of its contents because goal contents will be represented in natural language. Even if state-of-the-art NLP methods are adopted, there is no guarantee that the correct truth value of each goal can be identified, and at present, it is unclear that such an approach for the detection of conflicts among goals is sufficiently robust for practical use. There are also costs associated with the introduction of NLP. Meanwhile, *2-dip-dfs* is also capable of dealing with ambiguity by applying goal selection criteria such as *Cov*(g) or *Vcn*(g), the values of which are subjectively set by stakeholders after reviewing goal contents. However, *2-dip-dfs* has the merit of enabling conflict detection without requiring the application of NLP methods. This makes *2-dip-dfs* simpler compared with existing methods for detecting conflicts among goals. As NLP methods develop, it may be possible to incorporate them into *2-dip-dfs*. Both existing methods and *2-dip-dfs* share the need to introduce NLP methods to eliminate subjectivity and make goal selection precise.

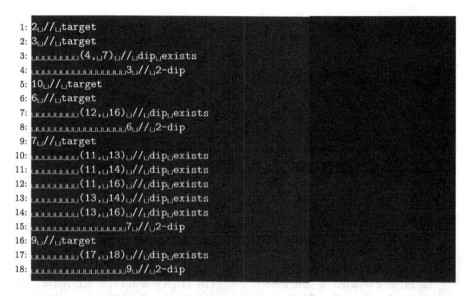

Table 2. Comparison between *Cov*(g) and *Vcn*(g) for each SGG_2^{OR} in Fig. 9. "A" means "Parent goal." "B" means "Pair between two child goals." "C" means "Comparison of *Cov*(g)." "D" means "Comparison of *Vcn*(g)." Note that each value of *Vcn*(g) is shown by rounding it to three decimal places if it cannot be divided. "E" means "Difference in priority." "F" means *2-dip*(g).

(a)

A	B	C	D	E	F
2	(4, 5)	$8 > 7$	$8 > 6$	No	$\perp$
	(4, 6)	$8 > 5$	$8 > -0.167$	No	
	(5, 6)	$7 > 5$	$6 > -0.167$	No	

(b)

A	B	C	D	E	F
3	(4, 5)	$7 > 6$	$8 > 6$	No	$\top$
	(4, 7)	$7 < 8$	$8 > 7.5$	Yes	
	(5, 7)	$6 < 8$	$6 < 7.5$	No	

(c)

A	B	C	D	E	F
6	(15, 12)	$6 > 5$	$6.667 > 3.667$	No	$\top$
	(15, 13)	$6 = 6$	$6.667 < 8.167$	No	
	(15, 16)	$6 > 3$	$6.667 > 6.167$	No	
	(12, 13)	$5 < 6$	$3.667 < 8.167$	No	
	(12, 16)	$5 > 3$	$3.667 < 6.167$	Yes	
	(13, 16)	$6 > 3$	$8.167 > 6.167$	No	

(d)

A	B	C	D	E	F
7	(13, 16)	$5 < 8$	$8.167 > 6.167$	Yes	$\top$
	(13, 14)	$5 < 8$	$8.167 > 6.167$	Yes	
	(13, 11)	$5 < 6$	$8.167 > 8$	Yes	
	(16, 14)	$8 = 8$	$6.167 = 6.167$	No	
	(16, 11)	$8 > 6$	$6.167 < 8$	Yes	
	(14, 11)	$8 > 6$	$6.167 < 8$	Yes	

(e)

A	B	C	D	E	F
9	(17, 18)	$6 < 8$	$3 > 2.5$	Yes	$\top$

(f)

A	B	C	D	E	F
10	(15, 12)	$6 > 4$	$6.667 > 3.667$	No	$\perp$

6 Conclusions and Future Work

In this study, we proposed a novel metric, *2-dip*(g), for detecting conflicts between two goal selection criteria based on a dfs algorithm. By applying the algorithm to a real goal graph, we demonstrated that *2-dip-dfs* functions as designed. However, the evaluation experiment also showed that *2-dip-dfs* should be extended to deal with the case of $SGG_2^{AND/OR}$. In future research, this issue will be addressed. Future studies should address two additional issues: (1) proof of the time complexity of *2-dip-dfs* and (2) generalization from *2-dip*(g) to *k-dip*(g) ($k \geq 2$) to enable conflicts among multiple goal selection criteria to be detected.

References

1. Aho, A.V., Hopcroft, J.E., Ullman, J.D.: Data Structures and Algorithms. Addison Wesley, Amsterdam (1983)
2. Alexander, I., Beus-Dukic, L.: Discovering Requirements. John Wiley & Sons, New York (2009)
3. Boehm, B.W.: Software Engineering Economics. Prentice Hall, Englewood Cliffs (1981)

4. Giorgini, P., Mylopoulos, J., Sebastiani, R.: Goal-oriented requirements analysis and reasoning in the tropos methodology. Eng. Appl. Artif. Intell. **18**, 159–171 (2005)
5. Horkoff, J., Yu, E.: Analyzing goal models - different approaches and how to choose among them. In: Proceedings of the 11th SIGAPP ACM Symposium on Applied Computing (SAC 2011), pp. 675–682 (2011)
6. Kaiya, H., Horai, H., Saeki, M.: AGORA: attributed goal-oriented requirements analysis method. In: Proceedings of the 10th Anniversary IEEE Joint International Conference on Requirements Engineering (RE 2002), pp. 13–22 (2002)
7. Kaiya, H., Shinbara, D., Kawano, J., Saeki, M.: Improving the detection of requirements discordances among stakeholders. Requirements Eng. **10**(4), 289–303 (2005)
8. van Lamsweerde, A., Darimont, R., Letier, E.: Managing conflicts in goal-driven requirements engineering. IEEE Trans. Softw. Eng. (TSE) **24**(11), 908–926 (1998). Special Issue on Managing Inconsistency in Software in Software Developtment
9. Mylopoulos, J.: Goal-oriented requirements engineering: Part II. In: Presentation Slides of the 14th IEEE International Requirements Engineering Conference (RE 2006) (2006). https://files.ifi.uzh.ch/rerg/arvo/events/RE06/ConferenceProgram/RE06_slides_Mylopoulos.pdf
10. Mylopoulos, J., Castro, J., Kolp, M.: Tropos: a framework for requirements-driven software development. In: Information Systems Engineering: State of the Art and Research Themes, pp. 261–273. Springer, Berlin (2000)
11. Nilsson, N.J.: Principles of Artificial Intelligence. Morgan Kaufmann Publishers, San Mateo (1980)
12. Robertson, S., Robertson, J. (eds.): Mastering the Requirements Process: Getting Requirements Right, 3rd edn. Addison-Wesley Professional, Boston (2012)
13. Saito, S., Yamamoto, S.: An attributed-based goal selection analysis method. J. Jpn Soc. Manage. Inf. **15**(3), 37–50 (2006). [in Japanese]
14. van Lamsweerde, A.: Requirements Engineering: From System Goals to UML Models to Software Specifications. John Wiley & Sons, Chichester (2009)
15. Yamamoto, S., Kaiya, H., Cox, K., Bleinstein, S.: Goal oriented requirements engineering trends and issues. IEICE Trans. Inf. Syst. **E89-D**(11), 2701–2711 (2006)
16. Yu, E., Giorgini, P., Maiden, N., Mylopoulos, J.: Social Modeling for Requirements Engineering. The MIT Press, Cambridge (2011)

Requirements Validation

Automated Support to Capture and Validate Security Requirements for Mobile Apps

Noorrezam Yusop[1], Massila Kamalrudin[1(✉)], Safiah Sidek[1], and John Grundy[2]

[1] Innovative Software System and Services Group, Universiti Teknikal Malaysia Melaka, Malacca City, Malaysia
p031320001@student.utem.edu.my, {massila, safiahsidek}@utem.edu.my

[2] School of Information Technology, Deakin University, Geelong, Australia
j.grundy@deakin.edu.au

Abstract. Mobile application usage has become widespread and significant as it allows interactions between people and services anywhere and anytime. However, issues related to security have become a major concern among mobile users as insecure applications may lead to security vulnerabilities that make them easily compromised by hackers. Thus, it is important for mobile application developers to validate security requirements of mobile apps at the earliest stage to prevent potential security problems. In this paper, we describe our automated approach and tool, called MobiMEReq that helps to capture and validate the security attributes requirements of mobile apps. We employed the concept of Test Driven Development (TDD) with a model-based testing strategy using Essential Use Cases (EUCs) and Essential User Interface (EUI) models. We also conducted an evaluation to compare the performance and correctness of our tool in various application domains. The results of the study showed that our tool is able to help requirements engineers to easily capture and validate security-related requirements of mobile applications.

Keywords: Security requirements · Security attributes · Validation · Test driven development · Mobile apps · Model based testing strategy · EUC · EUI

1 Introduction

Mobile phones have been used widely as they allow interactions between people and things anywhere and anytime. The use of mobile application is rapidly growing, especially in performing online transactions, such as online purchasing, flight booking and hotel booking. There are also a plethora of applications being developed to fulfil the needs of mobile users. However, many mobile application developers tend to ignore the security aspect of the application during the early stage of development, leading to malicious attacks and security breaches. It is also found that most of the requirements engineers fail to capture correct security related requirements during the elicitation phase as they face difficulties to understand the terms and knowledge of the security [1]. Further, the process of capturing correct and consistent requirements from

S.-W. Lee and T. Nakatani (Eds.): APRES 2016, CCIS 671, pp. 97–112, 2016.
DOI: 10.1007/978-981-10-3256-1_7

client-stakeholders is often difficult, time consuming and error prone [2, 3]. Therefore, there is a need for automation support to capture and validate security related requirements at the early stage of mobile application requirements engineering. In our previous work [4], we have conducted a user study to gauge requirements engineers' ability in capturing the security related requirements from a set of business requirements of a mobile application. The study found that the participants captured almost 60% incorrect security attributes for each of the requirements given. This result indicates that requirements engineers face difficulty to capture the security related requirements, especially in extracting the security attributes [4, 5]. Further, it was found that the longest time taken by the participants to extract the security attributes is more than 45 min, which means that more effort is needed to perform this task. These challenges have motivated us to: (1) develop an automated tool support for capturing and validating security requirements, and (2) evaluate the tool to demonstrate its ability to enhance the accuracy and usability for capturing and validating security requirements of mobile apps.

This paper describes the approach and an automated that captures and validates security requirements for mobile applications using Test Driven Development (TDD) with a model-based testing strategy using the Essential Use Cases (EUCs) and the Essential User Interface (EUI) models. We present background for this study, our prototype tool, and an experiment comparing its performance in extracting security attributes and validation from security requirements. Finally, we discuss implications and future work.

2 Background

2.1 Test Driven Development (TDD)

Test-driven development (TDD) is a development strategy that has been popularized by extreme programming [6]. The three important stages in TDD are (1) writing the test before adding the code, (2) writing the simplest code that passes the test, and (3) repeating this cycle until the software is matured. TDD also allows each of the requirements to be transformed to a test and it helps the engineers to think through the requirements or design before writing the functional code. TDD promotes a style of incremental development, where it identifies any behaviour that has been correctly implemented or remains undone. This approach enhances the analysis and the design of software as it allows the software to be tested at any time under automation [7].

2.2 Essential Use Cases (EUCs)

The EUC approach was defined by Constantine and Lockwood as a "structured narrative, expressed in a language of the application domain and of users, comprising a simplified, generalized, abstract, technology free and independent description of one task or interaction that is complete, meaningful, and well-defined from the point of view of users in a role or some roles in relation to a system and that embodies the purpose or intentions underlying the interaction" [8]. Its main objectives are to support better communication between the developers and stakeholders via a technology-free model and to assist better requirements capture. These objectives can be achieved by allowing only specific details relevant to the intended design to be captured [9]. EUCs enable users to ask fundamental questions, such as "what's really going on" and "what do we really need to do" without letting implementation decisions get in the way. These questions often lead to critical realizations that allow users to rethink, or reengineer the aspects of the overall business process. Figure 1 shows an example of natural language requirements (left) and an example of EUC (right) when capturing the requirements. The natural language requirements (highlighted) are shown on the left hand side.

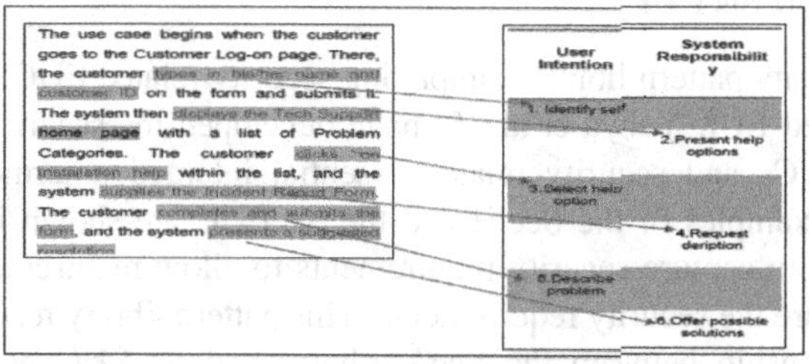

Fig. 1. Natural Language Requirements (left) and Essential Use Case (EUC) (right) [8, 10]

When capturing requirements from natural language text, the EUC model is found to be more suitable than the conventional UML use case. An equivalent EUC description is generally shorter and simpler than a conventional UML use case as it only comprises the essential steps (core requirements) of user's intrinsic interest. It contains the user's intentions and the system responsibilities to document the specific interaction without the need to describe the user's interface in detail. It is reported in [5] that EUCs are beneficial for capturing security requirements.

2.3 Essential User Interface (EUI)

EUI prototyping is a low fidelity prototyping approach [11]. It provides the general idea behind the UI instead of its exact details. Focusing on the requirements rather than the design, it represents UI requirements without the need for prototyping tools or widgets to draw the UI [12]. EUI prototyping extends from and works in tandem with the semi-formal representation of EUCs that also focuses on the users and their usage of the system, rather than the system features [13]. It thus helps to avoid clients and REs from being misled or confused by chaotic, evolving and distracting details. EUI also allows some explorations of the usability aspects of a system. Figure 2 shows examples of EUI prototype developed from EUC models.

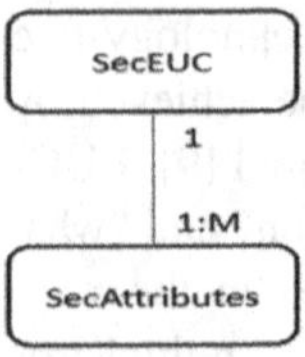

Fig. 2. The relationship between SecEUC model and SecAttrributes

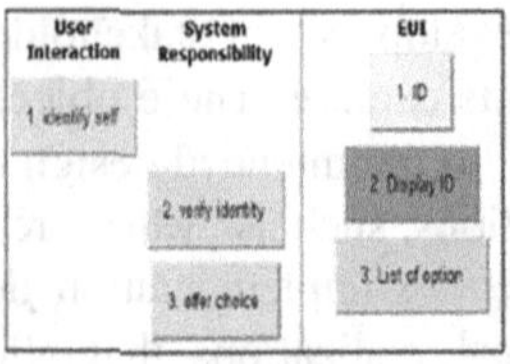

Fig. 3. Examples of EUI prototype from EUC models

Table 1. Example of SecEUC pattern libraries

SecEI	SecEUC	SecEUI	SecCtrl
Check username	Identify self	ID	Authentication
Check password			
Verify username			
Make payment	Make payment	Payment Type	Transaction
Complete payment form			

2.4 SecEUC and SecEUI

SecEUC is a security pattern library comprising security related EUC, while SecEUI is the security related EUI. Yahya et al. [5] have developed the security pattern library, called the SecEUCs and security related essential interaction termed as the SecEI pattern library. Examples of the SecEI and the SecEUC are shown in Table 1. They used EUC model to capture security requirements to allow requirements engineers to identify and capture the security requirements. This pattern library recognises that there is a direct relationship between the SecAttributes and SecEUC pattern library. As shown in Fig. 3, the relationship is one SecEUC to many SecAttributes (one to many). The main purpose for choosing the SecEUC and SecAttributes pattern library as well as their model is to conduct an in-depth analysis that could help to capture and validate security requirements from the business requirements. The SecEUC library patterns are based on EUCs generated from normal business requirements, while SecEI library patterns are based on the essential interactions found in security-related requirements.

Currently, both patterns only support the software/system development, but not the mobile application development. Further, the development of SecEUC patterns was adapted from the works of [14–16] and the identification of associated security elements are based on the definitions from the basic security services. Table 1 shows the example of SecEUC pattern libraries together with the relationship between many SecEI to one SecEUC and SecEUI. As shown in Table 1, the three SecEI identified as “check username”, “check password” “verify username” are related to one SecEUC “Identify self”, which is related to one SecEUI “ID”. Similarly, the two SecEI, which are “Make payment” and “Complete payment form” are related to one SecEUC “Make payment”, which is related to one SecEUI “Payment Type”.

3 Our Approach

The purpose of this study was to develop an approach and an automated tool to assist requirements engineers to automatically capture and validate the security-related requirements of mobile application. Our key research questions were: 1. Can an automated approach using TDD methodology with EUC and EUI models able to facilitate the extraction and validation of security attributes for security related requirements of mobile application? 2. Does the automation incorporated in the tool allow requirements engineers to quickly and accurately capture and validate the security attributes for security related requirements of mobile application? 3. How do the target users evaluate the usefulness of the automation tool in facilitating the extraction of security attributes and validation of the security requirements of mobile application?

Guided by these research questions, we developed an approach and a tool support, MobiMEReq, that automatically capture and validate the security requirements of mobile application. Here, the mobile application specific concerns and characteristics that reflected to this proposed approach are authentication, authorization and confidentiality. As shown in Fig. 4, our approach is depicted in the box labeled as A and our tool support is depicted in the box labeled as B. In this approach, we adopted the TDD methodology with a model-based testing strategy using EUCs and EUI models.

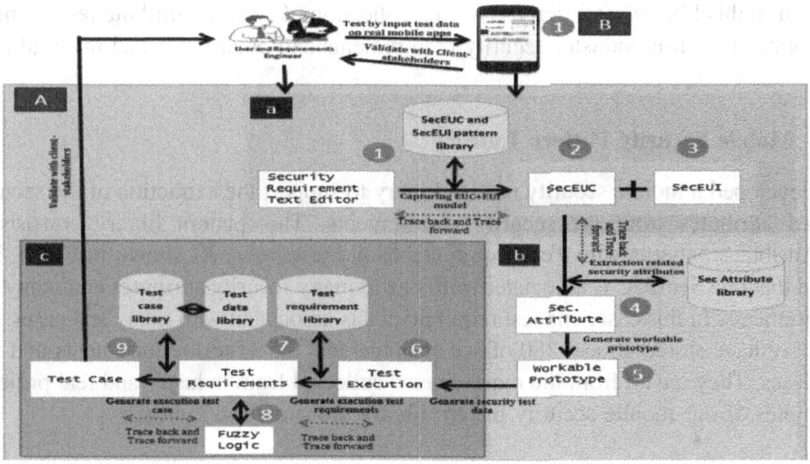

Fig. 4. Overview of our approach

As shown in Fig. 4, the approach is divided into three stages: capture the security requirements (a), generate the security attribute (b), and generate test (c). The first stage of our approach begins when the textual requirements are analyzed and traced to the EUCs patterns library for appropriate abstract interaction in a form of EUC model (1). Then, the SecEUC (2) and SecEUI (3) are derived from the generated EUC models based on the categorization of their attribute related to the security element as defined in the SecEUC pattern library. The second stage involves generating the security attributes. At this stage, each security attribute is generated from the SecEUC and SecEUI based on a defined security attribute library (4). Next, a workable prototype is generated to visualize the security requirements based on the generated SecEUI (5). This helps to validate the captured requirements with the textual captured requirements. Further, the security requirements are validated by the generated test that comprises test requirements (7) and test cases. This validation approach can also be done reversely from the generated and workable application prototype run in mobile to MobiMEReq

(1) where all the test components, security requirements attributes as well as EUC and EUI models are traced back. To realize our approach, we developed two pattern libraries, namely the mobile SecAttributes Pattern library and the Mobile Security Pattern Library. This approach also adopted the concept of fuzzy logic to prioritize the test requirements and test cases to validate the security requirements. The following section describes the development of these two libraries and the test prioritization.

3.1 Mobile SecAttributes Pattern Library

We developed the Mobile SecAttributes Pattern Library [17], consisting of SecEUC, SecEUI and related security attributes for mobile apps. Here, a security attribute is defined as any piece of information that may be associated with a controlled implicit entity or user for the purpose of implementing a security policy. Then, the SecEUC and SecEUI in this library were derived from a collection of security attribute requirements of mobile apps from industry security requirements and other published material [18].

3.2 Mobile Security Pattern Library

We developed a mobile security pattern library to support the extraction of the security related attributes from the security requirements. This pattern library consists of SecAttributes patterns with Test requirements and test cases. As shown in Fig. 5, it is found that one SecEUC is associated with one to many security attributes and many test requirements. In this case, one test requirements is associated with many test cases. For now, we have stored almost 280 of sec attribute and 185 of test requirements and 370 test cases. They are all from the requirements collected from industry and real projects. Examples of our mobile security pattern library are shown in Table 2.

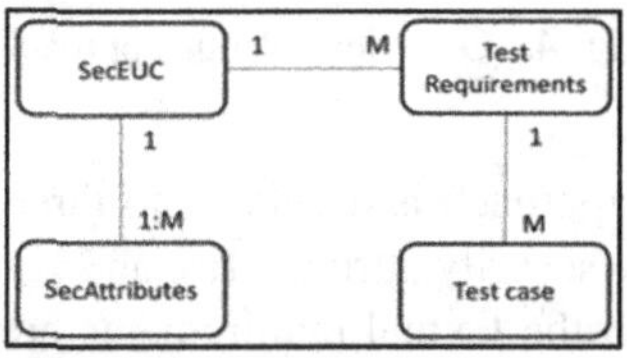

Fig. 5. The relationship of SecEUC with Sec attributes and test requirements

3.3 Test Prioritization Using Fuzzy Logic

Considering that a test case prioritization approach is able to improve the rate of faulty detection during the testing phase [19], we adopted the concept of test prioritization to prioritise the test requirements and test cases to validate the security requirements. In this case, the level of importance of the test case that runs during the test execution is identified according to the scale of high, medium and low, which has been validated by the experts. Further, the test case prioritization should be based on the captured requirements. In cases where the requirements are not prioritized, the test group is

Table 2. Sample of our SecAttributes pattern and test requirements

<table>
<tr><th rowspan="2">SecEUC</th><th rowspan="2">SecAttributes</th><th rowspan="2">SecCtrl</th><th rowspan="2">Test Requirements</th><th colspan="7">Test case</th></tr>
<tr><th>Test description</th><th>Pre-Condition</th><th>Test data</th><th>Steps</th><th>Expected Result</th><th>Actual result</th><th>Test Status</th></tr>
<tr><td rowspan="3">Make payment</td><td>Username</td><td rowspan="3">Authentication Transaction</td><td>validate that user can make payment</td><td>Validate that user can make payment if user account exist in database.</td><td>User require to login</td><td>Username='noorre zam' Password='noorre zam123' PaymentID='P112 23344'</td><td>1.Login 2.Key in Amount and click Submit button</td><td>User will be able to make payment to the system.</td><td>User successful to make payment to system.</td><td>PASS</td></tr>
<tr><td>Password</td><td rowspan="2">Validate that user cannot make payment</td><td rowspan="2">Validate that user cannot make payment if user account exist in database.</td><td rowspan="2">User require to login</td><td rowspan="2">Username='noorre zam' Password='noorre zam123' Amount='0' PaymentID='P112 23344'</td><td rowspan="2">1.Login 2.Key in Amount and click Submit button</td><td rowspan="2">User will be able to make payment to the system.</td><td rowspan="2">User unsuccessful to make payment to system.</td><td rowspan="2">FAIL</td></tr>
<tr><td>PaymentID</td></tr>
<tr><td rowspan="4">Verify TAC Code</td><td>Username</td><td rowspan="4">Authentication Authorization Transaction</td><td rowspan="2">Validate that user can verify TAC code</td><td rowspan="2">Validate that user can insert TAC Code if user account exist in database.</td><td rowspan="2">User require to login</td><td rowspan="2">Username='noorre zam' Password='noorre zam123' TAC Code ID='TAC112' PaymentID='P112 23344'</td><td rowspan="2">1.Login 2.Key in TAC CODE ID and click Submit button</td><td rowspan="2">User will be able to verify TAC Code to the system.</td><td rowspan="2">User successful to verify TAC Code to system.</td><td rowspan="2">PASS</td></tr>
<tr><td>Password</td></tr>
<tr><td>TAC Code ID</td><td rowspan="2">Validate that user cannot verify TAC code</td><td rowspan="2">Validate that user cannot insert TAC Code if user account exist in database.</td><td rowspan="2">User require to login</td><td rowspan="2">Username='noorre zam' Password='noorre zam123' TAC Code ID='TAC112' PaymentID='P112 23344'</td><td rowspan="2">1.Login 2.Key in TAC CODE ID and click Submit button</td><td rowspan="2">User will be able to verify TAC Code to the system.</td><td rowspan="2">User successful to verify TAC Code to system.</td><td rowspan="2">FAIL</td></tr>
<tr><td>Payment ID</td></tr>
</table>

required to propose the prioritization to clients for reviews. To do this, Fuzzy logic is adopted to our work. Fuzzy logic uses the 'uncertainty principle' that recognizes the use of an approximation rather than a fixed or exact value, which uses a range of true values instead of true or false value. It also uses a form of many-valued logic that has different ranges of membership between 0 and 1 defined by a fuzzy set [20]. This approach helps to automatically select the best test cases for each of the test requirements and helps to reduce the number of test cases [21]. Figure 6 shows the step-by-step procedure of applying the fuzzy logic to conduct test prioritization in our security requirements validation work. The algorithm as shown in Pseudo-code1 was applied to prioritize the test case from the generated test requirements.

The step by step to prioritize test case is shown in Fig. 6 and described below:

Step 1: To assign weight to the requirements, four prioritization factors (PF) are considered using fault severity as proposed by Kumar et al. [19]. We use this factor as an input parameter to the security requirements to embed in our fuzzy logic for target test requirements. These factors are: i. Business Value Measure (BVM): BVM is a measure, in which security requirements with the highest level of importance are the critical requirements to customer's business. The range of each requirement is from 1 (low) to 10 (high). Use cases can also be used to analyze the security requirements. ii. Project Change Volatility (PCV): PCV is based on how many times a customer modifies the project security requirements during the software development cycle. PCV is one of the criteria that help to assess the changes in the security requirement at the early stage after implementation of project is start. PCV is increase test efforts and the project is difficult to complete on time. iii. Development Complexity (DC): Each security requirement is analyzed based on the complexity of its implementation. Factors, such as development efforts, technology, environmental constraints and security requirement feasibility matrix are considered when measuring the complexity of the implementation. iv. Fault Proneness of Requirement (FPR): FPR is a measurement

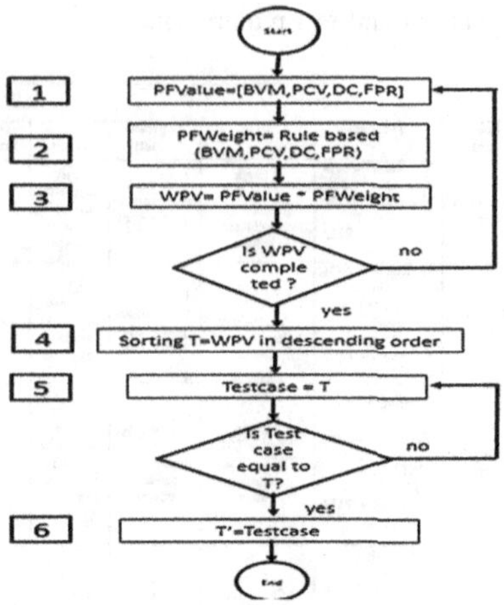

Fig. 6. Test requirement for prioritization test case algorithm Flow chart

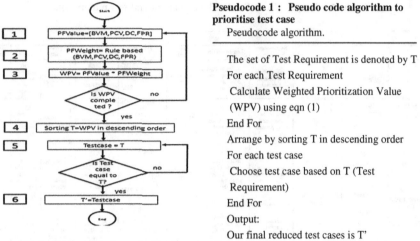

Pseudocode 1 : Pseudo code algorithm to prioritise test case

Pseudocode algorithm.

The set of Test Requirement is denoted by T
For each Test Requirement
 Calculate Weighted Prioritization Value (WPV) using eqn (1)
End For
Arrange by sorting T in descending order
For each test case
 Choose test case based on T (Test Requirement)
End For
Output:
Our final reduced test cases is T'

based on the error prone includes number or occurrence of in-house test failures found and also security requirements failures reported by the customer. Table 3 below show example an assignment low-high value for related test requirements based on four input factor.

Table 3. Low-high assignment test requirements

Range	**Low(1)-High (10)**			
Test Requirements	**BVM**	**PCV**	**DC**	**FPR**
validate that user can register to the system	2	10	3	6
validate that user can login to the system	2	5	5	5

Step 2: Figure 6 [2] shows the Priority Factor (PF) ruled based used as four input parameters to assign Priority Factor Weight. As shown in Table 4, there are: 4 parameters = BCM, PCV, DC, FPR. 3 memberships = Low, Medium, High; Rule-Based = 4 the power of 3 = 4 × 4 × 4 = 64 rules. Thus, based on the range applied in this study, 64 rules or less can be used. The pseudo-code algorithm for the rule based in our fuzzy algorithm is shown in Pseudo-code 2 below.

Step 3: Based on the four input parameters provided in the security requirements and the rule based, fuzzy inference system and defuzzification are then used to prioritize the test cases. The metric equation used for the prioritization of the test case is shown below.

Table 4. Weight prioritization value based on test requirements

	BCM	PCV	DC	FPR	WEIGHT
1	LOW	LOW	LOW	LOW	LOW
2	LOW	LOW	MEDIUM	LOW	LOW
3	LOW	LOW	HIGH	LOW	LOW
:	:	:	:	:	:
18	MEDIUM	HIGH	HIGH	HIGH	HIGH
:	:	:	:	:	:
63	HIGH	MEDIUM	HIGH	HIGH	HIGH
64	HIGH	HIGH	HIGH	HIGH	HIGH

Pseudocode 2 : Rule-based Pseudo code algorithm

If (BVM is low && PCV is low && DC is high && FPR is low) Then Weight is low
If (BVM is low && PCV is medium && DC is high && FPR is low) Then Weight is low
If (BVM is low && PCV is low && DC is low && FPR is low) Then Weight is low
If (BVM is high && PCV is high && DC is high && FPR is high) Then Weight is high
End IF

$$\text{WPV} = \left(\sum\nolimits_{pf=1}^{h} PFvalue * PFweight\right) \quad (1)$$

where, WPV is the weightage prioritization for each test case calculated based on the four input parameters. PF value is the value assigned to each test case. PF weight is the weight assigned for each input parameter. The weights are obtained from the fuzzy rules and the WPV is calculated from Eq. (1) which in turn gives the value of the prioritization order.

WP [TestReq 1] = 2 × 0.25 + 5 × 0.25 + 5 × 0.25 + 5 × 0.25 = 4, WP [TestReq 2] = 2 × 0.25 + 10 × 0.25 + 3 × 0.25 + 6 × 0.25 = 5.5,WP [TestReq 3] = 2 × 0.25 + 10 × 0.25 + 3 × 0.25 + 3 × 0.25 = 4.5

Step 4: Table 5 shows the sample results of the weight prioritization of test cases for the three functional requirements from the test requirements. Based on the calculation, the descending values of WPV are TestReq 2, TestReq 3, and Test Req 1.

Step 5: The selection number of test case from test requirements is selected as shown Fig. 7 based on the test requirements from ordering of WPV (Test Requirement).

Table 5. Samples of weight prioritization value based on test requirements

Factor / Test Requirements	TestReq1	TestReq2	TestReq3	Weight
BVM	2	2	2	0.25
PCV	5	10	10	0.25
DC	5	3	3	0.25
FPR	5	6	3	0.25
Weight Prioritization Value	4	5	4.5	1

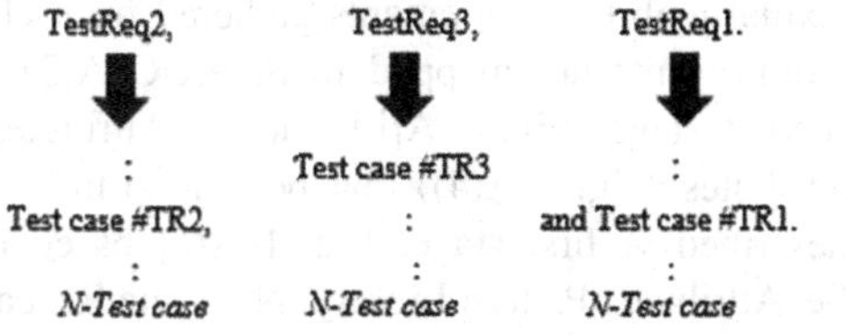

Fig. 7. Selection test case from test requirements

4 Implementation

We have developed a prototype tool called MobiMEReq to realise the approach of automatically capture and validation of security requirements as discussed in the previous section. This prototype tool is an extension of our earlier [14] tool that runs in both mobile and web applications. Figure 8 shows the tool usage based on the overview of our approach as shown in Fig. 4 and the role of the SecAttributes Pattern Library and the embedded fuzzy logic in validating the security requirements of mobile apps at the early stage of requirements validation. By implementing the TDD methodology and model based testing strategy, our approach is divided into two main parts: (1) Capturing security attributes from a set of mobile application requirements and (2) Validating the quality of security requirements.

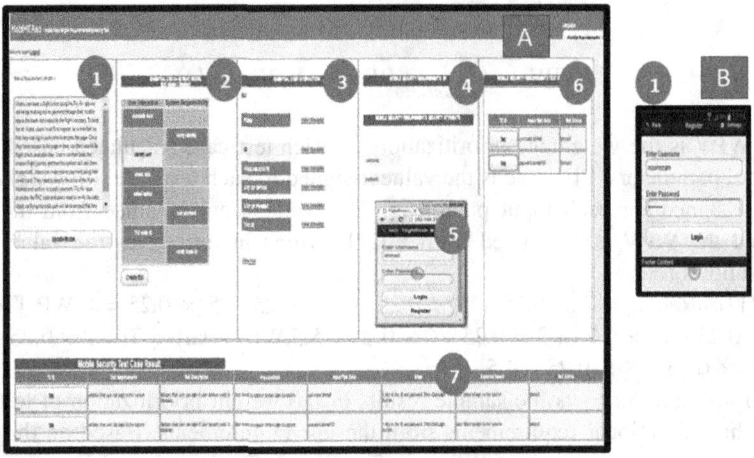

Fig. 8. Example of tool usage for integration security attributes and visualization tool

1. Capturing the Security Attributes [A]: Here, the Mobile SecAttributes Pattern Library is used by the tracing engine to analyse the textual mobile requirements and then match it to the set of abstract interactions of SecEUC (A1). This approach allows Requirements Engineers (REs) to capture the important security attributes from the textual mobile requirements gathered from client-stakeholders A(1). Then, the textual requirements are mapped to SecEUC A(2) and SecEUI model A(3). As shown in second stage, Fig. 4 A[b], the SecAttributes is generated to visualize the security attributes (Fig. 8 A(4)) that best fit to the generated SecEUC and SecEUI model as described at first stage, Fig. 4 A[a] based on the defined attributes in the Mobile SecAttributes Pattern Library. Next, the RE can visualise the security requirements as a form of workable rapid prototype model of the targeted mobile app [14].

2. Validating the quality of the security requirement consist of two parts: Part 1. Validation from MobiMEReq workable prototype [A].RE can visualise the security requirements in a form of workable rapid prototype model of the targeted mobile app.

This allows the client-stakeholder to check the quality of the generated security requirements in terms of its correctness and consistency with the original security related requirements. Further, to validate the correctness of the security requirements, RE can insert the test data into the workable prototype, Fig. 8 A(5). As shown Fig. 4, the results of the test review (labelled as A[c]) allow the execution of the test data and the display of the test status. RE can then perform the mobile Security Execution (SecExec) A[c](6) (Fig. 8, A (6)). Here, Test requirements A[c](7) are used based on the associated SecEUC. In order for RE to view the results of the test validation, the test cases A[c](9) are generated to visualise the validation based on the proposed fuzzy logic approach A[c](8). The results of the validation are also displayed as Pass or Fail as shown in Fig. 8, labelled as A(8) for each of the generated test data. Part 2. Validating the generated mobile apps with MobiMEReq [B]: Another utility provided by our MobiMEReq tool allows the generated mobile apps to be validated reversely to MobiMEReq. Both the RE and client-stakeholder can validate the functionality of the apps by inserting a random test data through their mobile apps and the associated components such as test cases A(7), test execution A(6), security attributes A(4) and both EUCs A(3) and EUI model A(2) and textual requirements A(1) as shown in Fig. 8. For large scale can view in this link[1].

5 Evaluation

Designed to be used by requirements engineers, we have conducted three studies to evaluate the accuracy and usability of our new automated tool in capturing and validating security requirements of mobile application. First, we conducted an accuracy test to evaluate the accuracy of the tool with three participants to check manually by applying a new set of security requirements. Secondly, we conducted a usability study with 50 undergraduate students to get their feedback based on four aspects, namely the usefulness, ease of use, ease of learning and satisfaction. Here a flight booking requirements that comprises security requirements such as login, payment and TAC code is used for the study. Finally, we requested two experts in the field of requirements engineering to test our tool and were later interviewed to collect their feedback.

5.1 Accuracy Test

We evaluated the accuracy of the tool by applying a new set of security requirements to check the false positive rate for the tool. False positive rate is used to calculate the ratio of the pass and fail for particular test. All of the related Test requirements were generated and they produced all correct false positive rate based on the prioritization of both test requirement and test case. The accuracy is measured based on the false-positive result of security requirements that provided by the tool compared to the sample answer prepared by the authors. The sample answer was first verified by the expert in the field of software engineering. Based on the results of the accuracy test

[1] https://drive.google.com/drive/folders/0B5QVa-tMkodvNXZnVlc3SGxMbkE.

shown in Table 6 (refer link to the accuracy test[2]), the MobiMEReq tool reported 100% correctness ratio based on the ten test requirements applied in the tool. This result indicates that the tool is able to facilitate requirements engineers to validate security requirements at the early stage of Software Development Life Cycle (SDLC).

Table 6. Sample of correctness of the automated validating tool

Test Requirements	Pre-Condition	Test data	Test Status (Pass/Fail)	Expected Result	Actual Result	No.Correct answers	No.Wrong answers
Validate that user can login	Register	Username='noorrezam' Password='noorrezam123'	P	User can login and welcome page is displayed	User can login and welcome page is displayed	1	0
Validate that user can login	Register	Username='noorrezam' Password='noorrezam123456'	F	User can login and welcome page is displayed	User cannot login and warning message is displayed	1	0
Validate that user can inserting TAC Code	Login	Username=noorrezam Password=noorrezam123 TAC Code='123456'	P	User can inserting TAC Code and warning message is displayed	User can insert TAC Code and warning message is displayed	1	0
Validate that user can inserting TAC Code	Login	Username=noorrezam Password=noorrezam123 TAC Code='123234'	F	User cannot inserting TAC Code and warning message is displayed	User cannot insert TAC Code and warning message is displayed	1	0
Correctness ratio						10 100%	0 0%

5.2 Usability Study

A survey to investigate the usability of the tool was also conducted with 50 participants. The questionnaire[3] was designed to gather participants' feedback regarding its usefulness, ease of use, ease of learning and satisfaction. In addition, the participants were requested to write their comments on the four aspects. Results of the study are shown in Fig. 9. With respect to usefulness, 80% of the participants felt that the tool is useful for capturing and validating security requirements. Only 1% of the respondents disagreed that the tool is useful and 19% of the participants were neutral. These results indicate that the majority of the participants agreed that the tool is useful. They also recommended that the tool need to be improved for better visualization so that users can view the summary result of test case at early of process of SDLC. With respect to ease of use, 77% of the participants agreed that the tool is easy to use, while only 5% disagreed with the statement. 18% of the respondents were indifference. These results also indicate that the majority of the participants agreed that the tool is easy to use. Among the comments given by the participants are that the tool helps users to simplify their use case and test requirement and they are proud to have a tool that can assist a new developer or requirements engineer to validate the requirements. In ease of learning, 78% of the participants agreed that the tool is easy to learn. 20% of the participants were neutral, while only 2% disagreed that the tool is easy to learn. The results indicate that the majority of the participants agreed that the tool is easy to learn. The participants further commented that they need the tool to overcome the difficulties of the learning process so that they can learn aspect of security requirements especially

[2] https://drive.google.com/drive/folders/0B5QVa-tMkodvb2ZuX3ROMzRGUWc.

[3] https://drive.google.com/drive/folders/0B5QVa-tMkodvYlVDQk9uelc0X1k

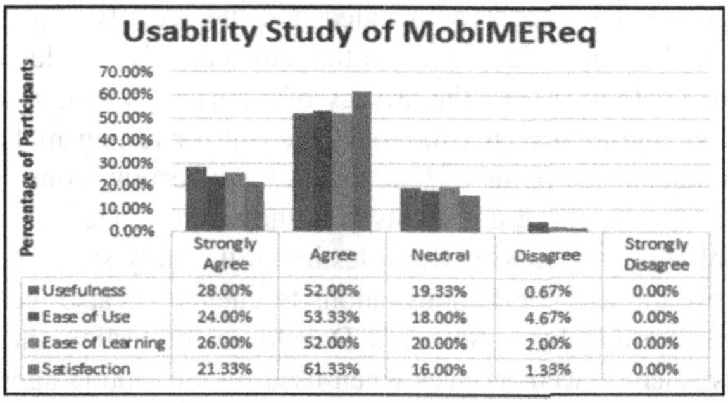

Fig. 9. Usability study for capture and validating security requirements on mobile app.

security attributes and validation process at early. The respondents were also requested to state their satisfaction of the tool. In this regard, 83% claimed that they were satisfied with the tool. 16% of the participants were indifference and only 1% were dissatisfied with the tool. The results indicate the majority of the participants were satisfied with our tool. However, we did not get 100% agreement from the participants that they are satisfied with the tool. Those who were dissatisfied with our tool commented that the tool needs to improve the process to complete testing. Based on the results of the survey, we can conclude that the tool is useful, user friendly and easy to learn. In general, most of the respondents were satisfied with the MobiMEReq tool.

5.3 Expert Review

We also carried out interviews with two experts in order to get their opinions regarding the usability of our prototype tool. We selected two experts in field of requirements engineering and quality assurance. They have between three to ten years working experience in the information technology (IT) industry. They are from MIMOS and IBM Malaysia and both are experienced in software requirements and test. Expert 1 has 8 years and Expert 2 is 3 years experience. Prior to the interview, we informed them the purpose of the interview and defined the different terminologies and definitions used in our interview questions to ensure the consistency of responses. We provided a brief description of our prototype tool and gave them the access to a link to explore the tool using samples of real-life requirements from their recent projects.

Based on the interviews, they agreed that MobiMEReq tool helps to reduce time and human effort in capturing and validating security requirements for mobile apps. They found the tool to be simple, easy to understand and learn. Nevertheless, they provided some constructive comments and feedbacks that are valuable for our research study. Expert 1 (E1) suggested that the tool with mobile apps should be applied for non-enterprise application as the tool can be used as a platform to test single mobile apps to validate user's needs or satisfaction rather than for the organization to access the mobile apps. This may help us to take less elapsed time to implement and validate in a small business application instead of cross of organization. In this case, the validation of security requirement with single test can be focus for whole process of business requirements. E1 also recommended that the security algorithm should be embedded in the tool for better testing execution efficiency. By producing this security algorithm, this may help to protect the privacy of the user's sensitive information and data. Expert 2 (E2) highlighted that the tool is useful for testing mobile apps during eliciting the requirements in terms of functionality and security considering that both RE and Software Developer are not involved at the testing stage in SDLC, in current practice. This tool allows them to perform testing at the early stage of SDLC. E2 felt that the RE can use the tool for demonstrations to clients. He also recommended that the tool could be applied by a Software Developer for testing their work in the back-end process, such as web services when used by the mobile apps.

6 Related Work

There have been many methods, approaches, techniques and tools used to validate security requirements for mobile applications. For example, Rhee et al. [22] discussed a methodology to test the Mobile Device Management (MDM) agent. The authors proposed the items and method to identify security requirements, the process and the real world test methods for android devices. Nevertheless, the security of the MDM agent to strengthen the security of the mobile devices needs improvements in validating mobile devices. Farhood et al. [23] proposed a data-centric model to protect all the vulnerabilities to prevent application and malware threats. They proposed a model to ensure the confidentiality, integrity and availability of data stored in mobile devices. However, they did not include validation for security requirement in their study. Gilbert et al. [24] applied the App Inspector, an automated security validation system that analyses application and generates report for potential security and privacy violation. The authors described the future need for making a more secured smartphones applications through automated validation. However, testing of the application does not cover validation in all aspects of security requirements. Gautam et al. proposed a novel secure data access and monitoring framework with data stored on employee's devices. The tool, referred as Concord identifies critical organizational data that have been accessed by the user's access and theft or loss of mobile device using the cryptography format. However, this tool does not provide a proper validation to confirm that the file is secure [25]. They claimed that security concerns have become very important especially when performing financial transactions, hence the assault vectors in this application need to be examined with due diligence [26]. In this respect, both the users and law enforcement agencies must tackle the issues of mobile security to reduce the risk of criminal misuse. Kamalrudin et al. [27, 28] developed a technique and toolset, MaramaAI to support requirements capture and consistency management using EUCs. This tool is supplemented by end-to-end rapid prototyping support. The tool uses EUC patterns to validate requirements consistency, completeness and correctness. However, it focuses on capturing language requirements only, hence the application of this tool in mobile application security requirements is beyond the scope of their work.

7 Summary

Security engineers need to validate security requirements for mobile application at an early stage of development. We have developed an automated validation approach and tool support called MobiMEReq for security requirements of mobile apps by adopting the idea of Test Driven Development (TDD) with model-based testing strategy using EUCs and EUIs prototype model. Evaluation of our prototype tool with real security examples and end users shows positive results.

Acknowledgement. We would like to thank Universiti Teknikal Malaysia Melaka and Sciencefund grant: 01-01-14-SF0106 and also Ministry of Education (MOE), MyBrain15 for support.

References

1. Schneider, K., Knauss, E., Houmb, S., Islam, S., Jurjens, J.: Enhancing security requirements engineering by organizational learning. Requirements Eng. **17**(1), 35–56 (2011)
2. Kamalrudin, M., Grundy, J.: Generating essential user interface prototypes to validate requirements. In: Proceedings of the 2011 26th IEEE/ACM International Conference on Automated Software Engineering, pp. 564–567 (2011)
3. Paja, E., Dalpiaz, F., Poggianella, M., Roberti, P.: STS-tool: socio-technical security requirements through social commitments. In: Proceeding of the Conference 21st IEEE International Requirements Engineering Conference (RE), pp. 331–332 (2012)
4. Yusop, N., Kamalrudin, M., Yusof, M.M., Sidek, S.: Challenges in eliciting security attributes for mobile application development. In: Proceeding of the Conference KSII The 7th International Conference on Internet (ICONI), Kuala Lumpur, Malaysia (2015)
5. Yahya, S., Kamalrudin, M., Safiah, S., Grundy, J.: Capturing security requirements using essential use cases (EUCs). In: First Asia Pacific Requirements Engineering Symposium, APRES 2014, pp. 16–30. Auckland, New Zealand, 28–29 April 2014
6. Paja, E., Dalpiaz, F., Poggianella, M., Roberti, P.: STS-tool: socio-technical security requirements through social commitments. In: Proceeding of the Conference 21st IEEE International Requirements Engineering Conference (RE), pp. 331–332 (2012)
7. SANS Institute, Determining the Role of the IA/Security Engineer, InfoSec Reading (2010)
8. Constantine, L.L., Lockwood, L.A.: Software for Use: A Practical Guide to the Models and Methods of Usage-Centered Design. Pearson Education, Upper Saddle River (1999)
9. Biddle, R., Noble, J., Tempero, E.: Essential use cases and responsibility in object oriented development. In: Proceeding of the 25th Australasian Computer Science Conference. Australian Computer Society, Inc., Chicago (2002). vol. 24(1), 7–16 (2002)
10. Constantine, L.L., Lockwood, A.D.L.: Structure and style in use cases for user interface design. In: Object Modeling and User Interface Design: Designing Interactive Systems. Addison-Wesley, Longman Publishing Co. Inc., pp. 245–279 (2001)
11. Ambler, S.W.: Essential (Low Fidelity) User Interface prototypes (2016). www.agilemodeling.com/artifacts/essentialUI.htm
12. Constantine, L.L., Lockwood, A.D.L.: Usage-centered software engineering: an agile approach to integrating users, user interfaces, and usability into software engineering practice. In: Proceeding of 25th International Conference on Software Engineering (ICSE 2003). IEEE Computer Society, Portland, Oregon (2003)

13. Ambler, S.W.: The Object Primer: Agile Model-Driven Development with UML 2.0, 3rd edn. Cambridge University Press, New York (2004)
14. Kamalrudin, M., Grundy, J., Hosking, J.: Tool support for essential use cases to better capture software requirements. In: Proceeding of IEEE/ACM International Conference on Automated Software Engineering, pp. 327–336 (2010)
15. Kamalrudin, M.: Automated software tool support for checking the inconsistency of requirements. In: 24th IEEE/ACM International Conference on Automated Software Engineering, ASE 2009. IEEE (2009)
16. Kamalrudin, M.: Automated support for consistency management and validation of requirements, Ph.D. thesis. The University of Auckland (2011)
17. Yusop, N., Kamalrudin, M., Sidek, S.: Capturing security requirements of mobile apps using MobiMEReq. In: Proceeding of 3rd Asia Pacific Conference on Advanced Research, Melbourne, Victoria, Australia (2016)
18. Yusop, N., Kamalrudin, M., Sidek, S.: Security requirements validation for mobile apps: a systematic literature review. Jurnal Teknologi (Sci. Eng.) **77**(33), 123–137 (2015)
19. Kumar, V.S., Kumar, M.: Test case prioritization using fault severity. Int. J. Comput. Sci. Technol. **1**, 67–71 (2010)
20. Novak, V., Perfilieva, I., Mockor, J.: Mathematical Principles of Fuzzy Logic. Kluwer Academic, Dodrecht (1999)
21. Bhasin, H., Gupta, S., Kathuria, M.: Implementation of regression testing using fuzzy logic. Int. J. Appl. Innov. Eng. Manage. **2**(4), (2013)
22. Rhee, K., Kim, H., Na, H.Y.: Security test methodology for an agent of a mobile device management system. Int. J. Secur. Appl. **6**(2), (2012)
23. Dezfouli, F.N., Deghantanha, A., Mahmood, R., Sani, N.F.M., Shamsuddin, S.: A data-centric model for smartphone security. IJACT **5**, 9–17 (2013)
24. Gilbert, P., Cun, B.: Vision: automated security validation of mobile apps at app markets. In: Proceeding of the 2nd International Workshop on Mobile Cloud Computing and Services (MCS 2011), pp. 21–26, New York, USA (2011)
25. Singaraju, G., Hoon, B.: Concord: a secure mobile data authorization framework for regulatory compliance. In: Proceeding of the 22nd Large Installation System Administration Conference (LISA 2008), pp. 91–102 (2008)
26. Ying, L., Dinglong, H., Haiyi, Z., Rau, P.: Users' perception of mobile information security. Hacker Journals White Papers. Computer Security Knowledge Base Portal (2007)
27. Kamalrudin, M., Grundy, J., Hosking, J.: Managing consistency between textual requirements. Abstract interactions and essential use cases. In: Proceeding of 2010 IEEE 34th Annual Computer Software and Applications Conference, pp. 327–336 (2010)
28. Kamalrudin, M., Grundy, J., Hosking, J.: Improving requirements quality using essential use case interaction patterns. In: Proceedings of 2011 International Conference Software Engineering, Honolulu, Hawaii, USA (2011)

A Template-Based Test-Authoring Tool to Write Quality Tests for Requirements Validation

Nor Aiza Moketar[1], Massila Kamalrudin[1(✉)], Safiah Sidek[1], Suriati Akmal[1], and Mark Robinson[2]

[1] Innovative Software System and Services Group, Universiti Teknikal Malaysia Melaka, Durian Tunggal, Malaysia
nor.aiza09@gmail.com,
{massila, safiahsidek}@utem.edu.my
[2] Fulgent Corporation, San Antonio, USA
marcos@fulgentcorp.com

Abstract. Requirements errors, such as incorrectness and incompleteness are the most common and difficult-to-fix defects in a software development project. In this paper, we describe the enhancement to our tool, TestMEReq with a template-based tests authoring to assist the requirements engineers in writing quality test requirements and test cases for requirements validation at the earliest stage of requirements engineering process. We embed an English language parser that checks the correctness of the test requirements and test cases written by the requirements engineers. In addition, prompt notification and highlight are also provided to visualize the errors and alert the requirements engineers. We conducted a user study to evaluate the usability of the tool and its effectiveness in helping novice requirements engineers to write quality tests.

Keywords: Requirements validation · Requirements-based testing · Test case · Essential use cases · Essential user interface

1 Introduction

Requirements errors are the most commonly found defect, comprising more than 80% of all product defects [1]. It is also the most expensive and difficult-to-fix defect at the later stage of software projects, wherein the cost can be more than one third of the total project cost [1]. Missing requirements, incorrect sentence structure or bad grammar and incorrect terms are found to be the most common problems when writing requirements [2]. In relation to this, it is important to ensure the correctness and completeness of the requirements statements so that they can be easily read and understood by both parties: requirements engineers and client-stakeholders. Good statements can also help to minimize misinterpretations and assumptions between requirements engineers and clients. Therefore, it is important to write correct and complete requirements at the earliest stage of requirements engineering process. Further, a mechanism such as template and guideline on writing good requirements and automated tool to validate

S.-W. Lee and T. Nakatani (Eds.): APRES 2016, CCIS 671, pp. 113–120, 2016.
DOI: 10.1007/978-981-10-3256-1_8

requirements are necessary to avoid inconsistency, incorrectness and incompleteness of the written requirements.

In our previous work, we have introduced a tool, called TestMEReq [3–5] to assist requirements engineers in validating the requirements captured from client-stakeholders. Our tool integrates the abstract models, namely the Essential Use Cases (EUC) and Essential User Interface (EUI) with requirements-based testing and rapid prototyping techniques. TestMEReq facilitates requirements validation process at the early stage by automatically generating abstract tests and a mock-up user interface (UI) prototype from the EUC and EUI models. It also provides a traceability function that allows users to trace back and forth between the textual requirements, the EUC and EUI models, the test requirements and the test cases to ensure their correctness, completeness, and consistency. In this paper, we discuss the enhancement of our tool, namely the provision of a template to author quality test as well as an automated mechanism to validate the written test requirements and test cases generated from the EUCs model. Following this section we provide some background of the study. Then, we present the description of our new approach followed by the usage example of the tool. Next, we presented our tool evaluation followed by the results and related study. We finally conclude our study and provide some overviews of our future works.

2 Background and Related Works

2.1 Test Requirements and Test Cases

Test requirements and test cases are the main components of our abstract tests generated from the semi-formalised abstract model called the Essential Use Cases (EUC) and Essential User Interface (EUI) model. Both test requirements and test cases are the high level abstraction of the requirements that does not contain any details of the test environment, test protocol or configuration for the test component. Both components are parts of the requirements, which are believed to contribute to the early testing in the requirements engineering phase. Here, the test requirements statements should be able to indicate the features of the requirements that need to be tested and validated. When doing the identification, it should include both normal and error conditions. In other words, the test cases should contain a set of input value, pre-condition of the testing and expected results that reflect the test requirements. However, authoring or writing quality test requirements and test cases are challenging as there is no specific guideline or rule and tool support to help requirements engineers to author quality test at the early stage of requirements engineering.

2.2 Template Based Approach

Template-based approach has been used in authoring by less expert/untrained/novice user to perform a task such as designing or programming [6]. It is intended to reduce a technical overhead or a skilled-set requirements for the design or development process, and to increase the productivity efficiency as well as promote reusability. Template-based tool generally provides the facility to create, edit, review, test and configure requirements. It also provides guideline, advice/feedback on/for a particular artifact. Template-based approach has been applied in various application domains such as education [7], programming [8], and multimedia design [6]. S. Davis, P. Bogen, L. Cifuentes et al. [7] have presented a Template-Based Authoring Tool (TBAT) for creating pedagogical sound Web-based activities in the form of Walden's Path. It aids the authors to create paths based on specialized academic as well as non-academic template. The evaluation results showed that TBAT reduces the preparation time for authoring of paths, thus leading to increased productivity.

R. Palski, E. Phipps, A. Salinger [8] have presented an approach for incorporating embedded simulation and analysis capabilities in complex simulation codes through template-based generic programming. This approach relies on templating and operator overloading within the C++ language. This approach is complex and requires users to have high familiarity with the template. However, it allows the programmer to ignore most of the part of the template.

N. Ali, J. Hosking, J. Huh, J. Grundy [9, 10] have introduced the Marama Critic Definer which provides support to end-user and tool designer for critic authoring and configuration task. Although it only supports simple design critic construction, this tool assists novice designer/developer to easily design critics by providing visual critic authoring template.

Based on our investigation, we found that template-based approach has benefited various application domains that help users to be more productive and efficient. However, study or works that discuss template for assisting novice requirements engineers to write quality test especially for requirements validation process are almost none-existence.

3 Our Approach

This paper will discuss the enhancement of our work called TestMEReq, with a template-based test authoring to assist the requirements engineers in writing quality test requirements and test cases. Figure 1 shows the overall overview of our previous work in capturing requirements and generating test from the EUC and EUI models (labeled as 1–6) and our new work on test-authoring template (labeled as 7), as highlighted in the grey box. Our test authoring templates consist of the templates for writing test requirements and test cases. The main objectives of these templates are three-folds: (1) to assist novice requirements engineers to write quality test, (2) to enhance the productivity and efficiency of the untrained requirements engineers in writing test, and (3) to enhance the scalability of our requirements and test pattern libraries. This template has three main functions, which are to create, to update and to delete the test requirements (TR) or test cases (TC). In our work, the TR and TC are considered as a part of the requirements specification where it is used for front-end testing to validate the user's requirements. As described previously, our tool can generate the relevant TR from the EUC model, which is supported by our TR pattern library. Similarly, the TC can be automatically generated from the TR. Our template allows the user to create/add new TR/TC if it is not generated automatically. This may be the case when the TR/TC is not defined in our pattern library or it is linked to another EUC model. Thus, our template allows the requirements engineers to create new TR/TC to the EUC model, if required.

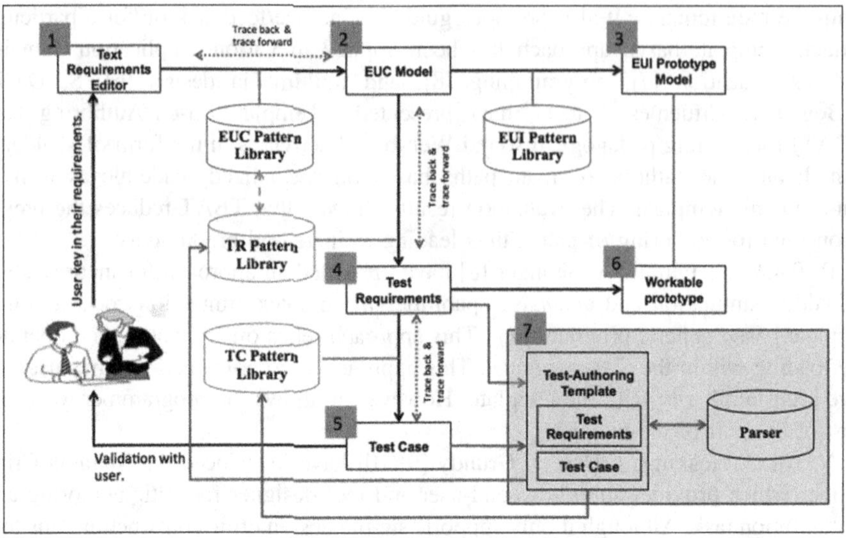

Fig. 1. An overview of our approach

In order to develop the test-authoring template, we have embedded a natural language parser of English language to ensure the correctness and accuracy of the sentence structure of the new TR and the TC provided by the REs. The parser helps to ensure that the user follows the correct sentence structure from our TR pattern library. This function helps to ensure the uniformity of the sentence in the pattern library. The sentence structure of our TR follows the key textual phrase as shown in Fig. 2. The parser ensures that the RE uses the right terms for the TR and that the sentences reflect the objectives/goals of the requirements. For this, the user must use an action verbs and words such as "Validate that…", "Verify that…" and "Test that…" in the test requirements statements. Some examples of sentences that follow the sentence structure of our TRs are:

1. Validate (VB) that (Art) user (NN) can (MD) login (VB) with (Prep) valid (Adj) user name and password (NN).
2. Validate (VB) that (Art) user (NN) can (MD) withdraw (VB) the correct (Adj) amount (NN).

Our tool also features prompt notification and feedback as well as highlights errors to alert requirements engineers to any defects found in the TR and TC. The tool is flexible as it allows users to ignore a notification. This may be helpful if the user thinks that the addition is relevant to the test requirements and test cases, and the newly added test requirements and test cases can be reviewed later by the REs. Further, the test-authoring template enhances the scalability of our test requirements pattern library by allowing requirements engineers to insert relevant test requirements and test cases for various software subject domains.

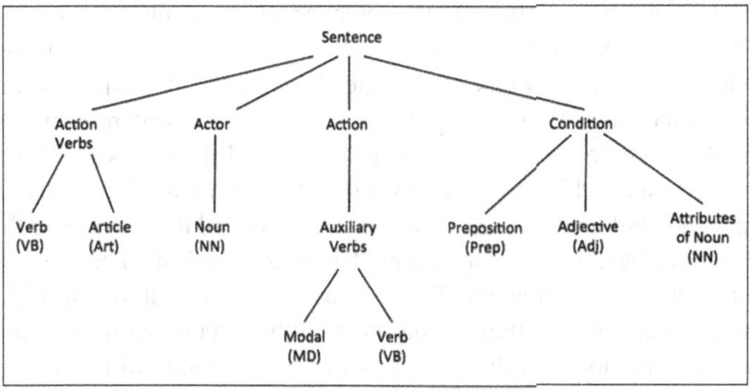

Fig. 2. Tree Structure for key textual phrase

4 Usage Example

To demonstrate the key features of our tool that is the test-authoring template, we use the user persona of a requirement engineer who would like to add a new test using the template of automotive-based requirements. In this case, Lily, a requirements engineer would like to add new test requirements for the "Broadcast Error" use case of the "Automotive On-board Diagnostic System" adapted from [11]. To do this, as shown in Fig. 3-label 1, she selects the "Broadcast Error" from the list of EUC models. Then, she clicks on one of the test requirements listed under the EUC model of "Broadcast Error", in which she will be directed to the test requirements template editor page, as shown in Fig. 3-label 2. Here, she can choose whether to edit or add new test requirements by choosing the appropriate option from the dropdown list (A). After that, she needs to key

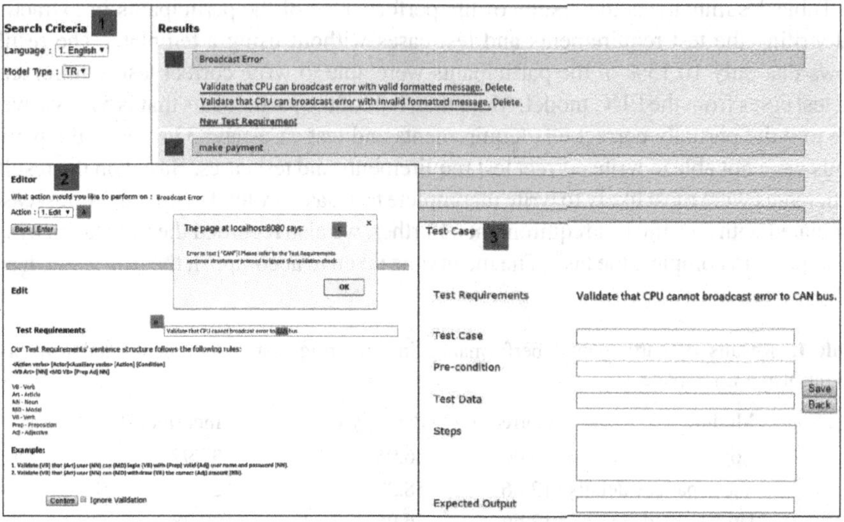

Fig. 3. Test requirements template editor

in the new test requirements following the sentence structure rules (B). In this example, she entered a new test requirements for "broadcast error", which is "Validate that CPU cannot broadcast error to CAN bus". Once she clicks on the "Confirm" button, the tool prompts an error message, highlighting the error found in the sentence. In this example, the tool cannot recognize the word "CAN" provided by Lily as it is the abbreviation of "Controller Area Network", which is considered as a noun. Here, Lily can decide whether to proceed with the new test requirements or amend it according to the sentence structure rule. Once the new test requirement has been added, she needs to add new test cases related to the test requirement. The test case template is shown in Fig. 3-label 3. The test case template ensures that she has entered the important component of the test case such as pre-condition, test data, steps and expected output of the test.

5 Evaluation and Result

We have conducted a user study to evaluate the prototype of our template-based test-authoring tool with 30 undergraduate students from software validation and verification course. The user study was conducted to evaluate the participant's performance in writing/authoring test requirements and test cases manually without any template/guideline. The participants were provided with a sample requirement of login function. They were asked to explore the tool with the sample requirements. They were informed that they would be observed and were encouraged to speak aloud their views of the tool while completing the task. The purpose of the observation was to identify problems and misconceptions faced by the participants when using the tool. The verbal evaluation of the tools provided us with the users' spontaneous responses and suggestions for improvement as they used the tool. After giving specific time to explore our tool, they were asked to complete a survey questionnaire regarding the usefulness, ease of use, ease of learning and satisfaction of the tool based on a five-level Likert scale.

Table 1 summarizes the results of the performance of the participants in extracting and writing the test requirements and test cases without using a template. The results shows that only 10.13% of the participants were able to write correct test requirements and test cases from the EUC model. More than half of the participants that is 57.38% were able to write partially correct test requirements and test cases and 32.49% of the participants were not able to write correct test requirements and test cases. Based on the results, participants were most likely to write incomplete tests, as they tend to miss a few test cases associated with specific test requirements. Further, we also recorded the time taken by the participants to complete the task. The mean time taken to accomplish the task was 1 h and

Table 1. Results of participants' performance in authoring test requirements and test cases without using a template

Module	Correct (%)	Partially correct (%)	Incorrect (%)
Login	5.06	56.96	37.97
View patient details	12.66	58.23	29.11
Update medication	12.66	56.96	30.38
Mean	10.13	57.38	32.49

15 min (75 min). The shortest time taken to complete the task was 55 min. This study also shows that it is time consuming and tedious to manually write correct, complete and consistent test requirements and test cases from the requirements model.

Figure 4 presents the results of the usability study. In term of the usefulness of the tool, 50% of the participants agreed and 40% strongly agreed that the tool was a useful tool to assist them to write test requirements and test cases. They also found that 38.9% of the participants felt that the tool was very easy, and 45.6% felt that the tool was always easy to use. In terms of the ease of learning, 33.3% of the participants claimed that it was very easy to learn since the flow and the interface design of tool is simple and user-friendly. Additionally, 21.1% of the participants were very satisfied, and 53.3% of the participants were always satisfied with the tool as no special technical skills to write accurate test requirements and test cases. Overall, the usability results show that our prototype tool is useful, easy to use and learn. Users also expressed their high level of satisfaction when using the tool.

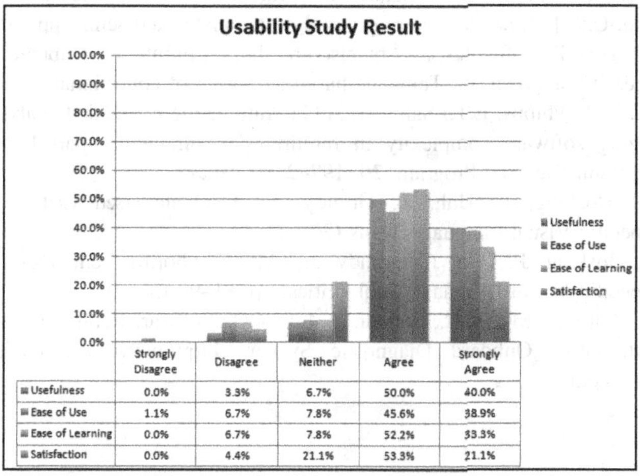

	Strongly Disagree	Disagree	Neither	Agree	Strongly Agree
Usefulness	0.0%	3.3%	6.7%	50.0%	40.0%
Ease of Use	1.1%	6.7%	7.8%	45.6%	38.9%
Ease of Learning	0.0%	6.7%	7.8%	52.2%	33.3%
Satisfaction	0.0%	4.4%	21.1%	53.3%	21.1%

Fig. 4. Usability study result

6 Conclusion and Future Works

We have enhanced our tool with template-based test authoring to help requirements engineers to write quality test requirements and test cases. This template can avoid common issues in writing requirements, such as incorrect sentence structure, bad grammar and incorrect terms. For future work, we plan to conduct industrial case studies to further evaluate the usability of our tool.

Acknowledgment. We would like to thank Ministry of Education (MOE) Malaysia, Universiti Teknologi Mara (UiTM), Fulgent Corporation, USA and FRGS: FRGS/1/2014/TK01//FKP/02/F00230 for funding this research. We also thank Noorrezam and Luqman for their assistance in this study.

References

1. Young, R.R.: Effective Requirements Practice. Addison-Wesley Information Technology Series (2001)
2. Hooks, I.: Writing good requirements (A Requirements Working Group Information Report). In: Proceedings of the Third International Symposium of the NCOSE (1993)
3. Moketar, N.A., Kamalrudin, M., Sidek, S., Robinson, M., Grundy, J.: TestMEReq: generating abstract tests for requirements validation. In: Proceedings of the 3rd International Workshop on Software Engineering Research and Industrial Practice - SER&IP 2016, pp. 39–45. ACM Press, New York (2016)
4. Moketar, N.A., Kamalrudin, M., Sidek, S., Robinson, M., Grundy, J.: An automated collaborative requirements engineering tool for better validation of requirements, pp. 864–869 (2016)
5. Kamalrudin, M., Moketar, N.A., Grundy, J., Hosking, J.: Automatic acceptance test case generation from essential use cases. In: 13th International Conference on Intelligent Software Methodologies, Tools and Techniques, pp. 246–255. IOS Press (2014)
6. de Albuquerque Azevedo, R.G., Santos, R.C.M., Araújo, E.C., Soares, L.F.G., de Salles Soares Neto, C.: Multimedia authoring based on templates and semi-, pp. 205–214 (2013)
7. Davis, S., Bogen, P., Cifuentes, L., Francisco-revilla, L., Furuta, R., Hubbard, T., Karadkar, U.P., Pogue, D., Shipman, F.: Template-based authoring of educational artifacts (2006)
8. Pawlowski, R.P., Phipps, E.T., Salinger, A.G.: Automating embedded analysis capabilities and managing software complexity in multiphysics simulation, part I: template-based generic programming. Sci. Program. **20**, 197–219 (2012)
9. Ali, N.M., Hosking, J., Huh, J., Grundy, J.: Template-based Critic Authoring for Domain-Specific Visual Language Tools (2009)
10. Ali, N.M., Hosking, J., Huh, J., Grundy, J.: Critic Authoring Templates for Specifying Domain-Specific Visual Language Tool Critics, pp. 81–90 (2009)
11. Fielstra, R., Adhi, R., Marsh, C., Bodnar, L., Lotus, C.: Software Requirements Specification (SRS) Automotive Onboard Diagnostic System. http://www.cse.msu.edu/~435diag3/DCubed_SRS.pdf

Requirements Analysis

A Digital Business Analysis Method

A Business Requirements Engineering Process by the Cooperation of Management, Business Operation and IT Department

Ichiro Yamaguchi[1], Masanobu Furukawa[2], Mikio Aoyama[3(✉)], and Yasuhiro Kikushima[3]

[1] TOKYO GAS i NET CORP, 2-4-1 Hamamatsu-cho, Minato-ku, Tokyo 105-6013, Japan
ichiro@tg-inet.co.jp

[2] Japan Exchange Group, Inc., 2-1 Nihombashi Kabuto-cho, Chuo-ku, Tokyo 103-8224, Japan
m-furukawa@jpx.co.jp

[3] Nanzan University, 18 Yamazato, Showa-ku, Nagoya 466-8673, Japan
mikio.aoyama@nifty.com, y-kiku@ark.ocn.ne.jp

Abstract. Business innovation is driven by information technology. Therefore, business systems are not separable from information technology. This article proposes a DBA (Digital Business Analysis) method, which is intended to create requirements for business system and information system in an integrated way. The DBA method is intended to collaborate both business and IT departments, and to align the digital business system with the management strategy.

Keywords: Requirements engineering · Business requirements · Business analysis · Business Process Management · Digital business · Collaborative requirements engineering

1 Introduction

Business innovation is driven by information technology. Therefore, business systems are not separable from information technology. In this article, we call digital business system, or simply digital business, as the business systems driven by information systems [10]. It is necessary to systematically develop requirements of business systems together with information systems, provided that the business system meets the management strategy.

While requirements engineering for software systems has been evolved and a body of knowledge has been formed [1, 7], there are similar but different disciplines in the arena of the requirements engineering for business systems [2], including BA (Business Analysis) [4, 8], and BPM (Business Process Management) [3, 6]. It is necessary to establish a concrete method to analyze digital business systems.

S.-W. Lee and T. Nakatani (Eds.): APRES 2016, CCIS 671, pp. 123–131, 2016.
DOI: 10.1007/978-981-10-3256-1_9

Thus, this article tries to answer the following research questions:

(1) What is an appropriate process model to analyze digital business system, and,
(2) What is missing in the conventional business analysis model, such BABOK, in the analysis process of digital business process';

2 Related Works

(1) RE for Business Systems
Both ISO/IEC/IEE 29148 [5] and REBOK [1, 7] define three layers of scope of requirements, that is, business systems, information systems, and software systems. They also define a set of processes for engineering requirements. However, the processes tend to focus on software requirements, and do not address much on business systems.
(2) BA (Business Analysis) and BABOK (Business Analysis Body Of Knowledge)
BA is a discipline to analyze and structure business requirements [8]. As a BOK, BABOK defines six processes to engineer business requirements [4]. However, the BABOK focuses on a business system, and engineering information systems is a separate issue. Therefore, BABOK has a limitation in engineering digital business systems.
(3) BPM (Business Process Management)
BPM is a comprehensive discipline of designing and managing business processes [6]. BPM CBOK (Common Body Of Knowledge) defines five major processes [3]. Therefore, the scope of BPM is beyond requirements engineering. However, by its definition, BPM focuses on business process first, and information system is considered as a way to support the business process.

3 Approach

Our approach is illustrated in Fig. 1. Conventional business analysis is based on the top-down approach, where the business strategy is input to the business analysis. The requirements specifications of a future business system, that is a business system to be, is elaborated. Then, the requirements specifications are documented for supporting the

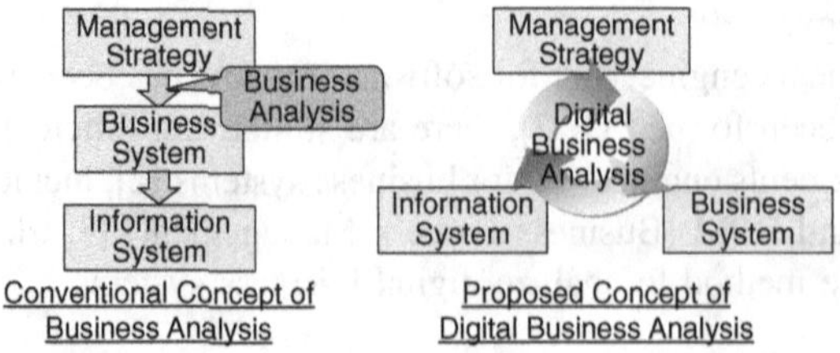

Fig. 1. Approach

future business system. The business analysis is generally conducted by either the staffs of IT department or so called business analysts.

The authors view that the analysis of business system is not separable from that of information system. Furthermore, both business system and information system are responsible to meet the management strategy, and should be validated together to the business strategy.

Based on the triangular structure in Fig. 1, the analysis process of digital business should be iterative in order to adapt to the rapid change of the business environment and IT in an agile way.

We also emphasize the active involvement of business department in the analysis of business system in collaboration with IT department. This is critical to elicit "true" business requirements since the only business department knows them.

4 Digital Business Analysis Method

4.1 DBA Framework

Figure 2 illustrates the framework of DBA (Digital Business Analysis). Three integral parts of the framework include DBR (Digital Business Requirements), DBA Process (Digital Business Analysis Process), and Collaboration Model.

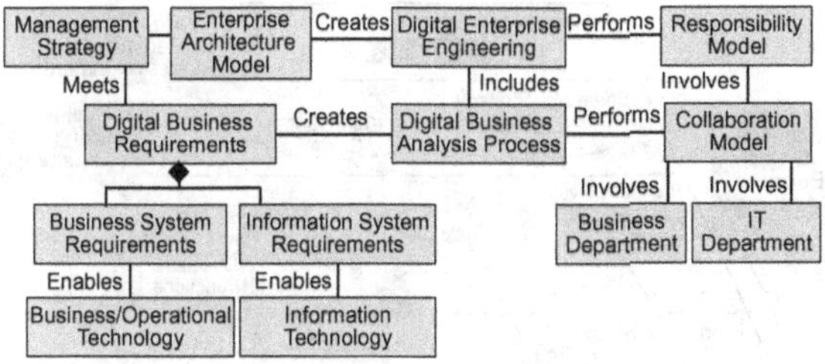

Fig. 2. DBA framework

Here, we refer *digital business* as a class of business systems driven and defined by information systems [10]. Therefore, we assume DBR consists of business system requirements and information system requirements, which are respectively enabled by business/operational technology and IT, in an integrated way. Management strategy is unfolded to the goals of DBR, which is determined by the corporate management based on the business strategy and/or market strategy.

We assume the DBA process is a part of digital enterprise engineering. It includes enterprise architecture, governance structure, and so on. Although the framework of digital enterprise engineering is interesting, this article focuses on the DBA process.

The Enterprise Architecture Model illustrates a holistic view of the architecture of an enterprise in which the Management Strategy, Enterprise Strategy and Digital Business System are structured.

The Collaboration Model presents how the business department and IT department work together in the DBA process. Similarly, Responsibility Model illustrates a big picture of responsibility of departments in the digital enterprise engineering, and position of the DBA process together with collaborating departments.

4.2 Enterprise Architecture Model

Figure 3 illustrates the scope of digital business systems in the Enterprise Architecture Model. The Enterprise Architecture Model represents three-layered architecture of enterprise systems which consists of:

(1) Management strategy,
(2) Enterprise strategy, and
(3) Digital business system.

The digital business system consists of business system and information system in an integrated way.

The relationship between layers is modeled by either goal and sub-goal, or goal and function(s) in order to meet the business goals [9]. Through the relationships, it is possible to validate the alignment between the management strategy and the digital business system.

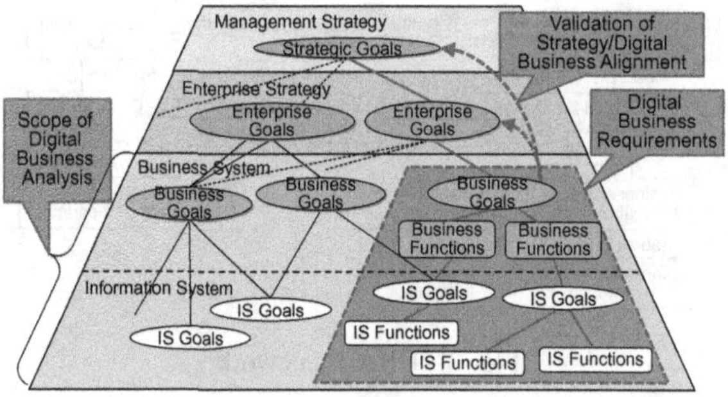

Fig. 3. Enterprise Architecture Model

4.3 Responsibility Model

Figure 4 illustrates Responsibility Model, which guides how each organization in an enterprise works together to elaborate a series of specifications in an iterative way.

Unlike conventional business planning, which is done in a top down way, the proposed DBA process strongly encourages collaboration between business department and IT department. We expect two major effects of the collaboration:

(1) Business department is an owner of the digital business system, and is responsible for the operation and business performance of the new business system, and,
(2) IT department needs to closely work together with business department in order to realize "true" requirements with agility.

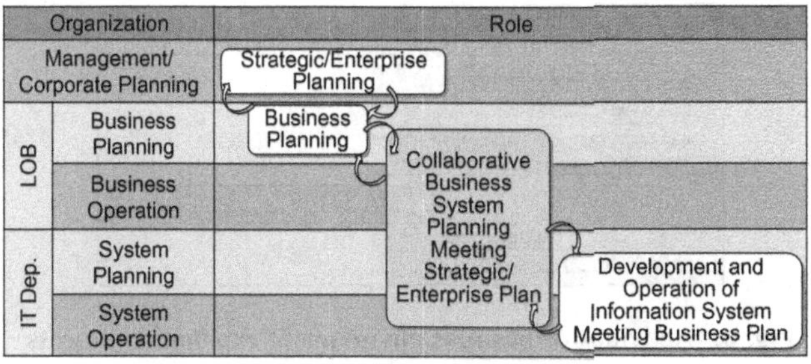

Fig. 4. Responsibility Model

4.4 DBA Process Model

Figure 5 illustrates the DBA process model, which consists of ten processes in the following three stages. We defined the details of each process, called activities, and identified 34 activities.

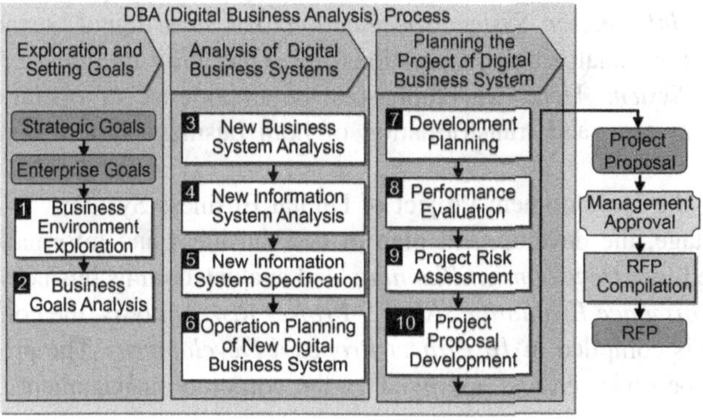

Fig. 5. DBA process model

As an example, the activities of the process "Business Environment Exploration" is illustrated in Fig. 6. In the figure, each activity is located on the two swim lanes, conducted by the business department and IT department. Therefore, the position of activity indicates the responsibility of two departments in the execution of the task.

(1) Exploration and Setting Goals
The proposed DBA process is business-goal-driven [9]. The business goals are a set of goals which the digital business system needs to meet. The goals can be identified from the strategic and enterprise goals, and the results of *1 Business Environment Exploration.* In *2 Business Goal Analysis*, the business goals are structured from the strategic goals to the tactical goals. The goals drive the subsequent processes of digital business analysis.

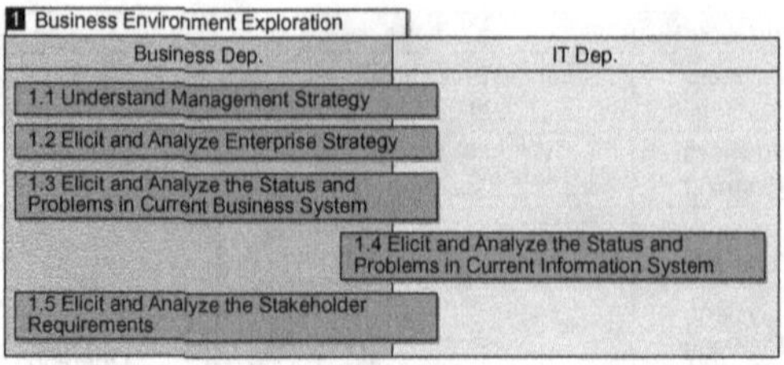

Fig. 6. Activities of business environment exploration process

(2) Analysis of Digital Business Systems
This stage is an integral part of DBA process, and consists of four processes. The first two processes, *3 New Business System Analysis*, and *4 New Information System Analysis*, can be evoked concurrently, or iteratively, although they appear to be sequential in Fig. 5. The execution patterns of the two processes depend on the design context, whether you may start thinking from business, information system, or both, since business system and information system are inseparable in digital business systems. The results of the analysis are compiled as a specification in *5 New Information System Specification.* The new digital business system is specified after analyzing its operations in *6 Operation Planning of New Digital Business System.* Here, we emphasize the importance of operation of digital business system, and transformation of both business system and information system.

(3) Planning the Development Project of Digital Business Systems
In this stage, the development plan of new business and information system is developed in *7 Development Planning*, and evaluated on its business performance in *8 Performance Evaluation.* After *9 Project Risk Analysis*, the complete project proposal is compiled in *10 Project Proposal Development.* The project proposal needs to be reviewed and approved by the corporate management.

4.5 Collaboration Model

Figure 7 illustrates a model of organizational collaboration between business department and IT department. In each department, we identified two business units, planning and operation under business department, and development and operation under IT department. The number of "+" indicates the degree of involvement, "++" denotes the unit of leading the process, and "+" denotes units collaboration with the leading unit.

This collaboration model is particularly important in the digital business analysis. Furthermore, we strongly recommend that business operation unit should lead the most of the processes, including *1* to *3*, *6*, and *8* to *10* since the only business operation unit elicits and knows the real requirements in order to create new business value by the new business system. The operation department is also responsible to evaluate the value of the new business system, that is, the *validation* of the new business system.

Process	Business Department		IT Department	
	Planning	Operation	Development	Operation
1. Business Environment Exploration		++	+	+
2. Business Goals Analysis		++	+	
3. New Business System Analysis		++	+	
4. New Information System Analysis		+	++	
5. New Information System Specification		+	++	+
6. New Digital Business System Operational Specification		++	+	+
7. Development Planning	+	+	++	+
8. Performance Evaluation	+	++	+	+
9. Project Risk Analysis	+	++	+	+
10. Project Proposal Development	+	++	+	

Legend ++: Lead, +: Collaborator

Fig. 7. DBA collaboration model

5 Evaluation and Discussion

The authors evaluate and discuss the proposed DBA process with respect to the two research questions, RQ1 and RQ2.

5.1 RQ1: Evaluation of the Proposed DBA Process

The authors evaluated the proposed DBA process with the following two perspectives.

(1) Comparison of the DBA process with the business analysis processes in practice
The authors compared the proposed DAB process with current business analysis processes conducted in two companies. The two companies are large-scale enterprises providing infrastructure services in Japan. By the comparison, we found that the business analysis processes of two companies lack some activities of the proposed DBA process. Although the example is limited, the proposed process is considered to be more complete. The proposed DBA process is practically applicable since it provides enough details in the activities.

(2) Feasibility analysis with an example of business transformation
The authors also applied the proposed DBA process to an example of digital business transformation. In the example, the authors found that it is particularly important to analyze both business process and IT systems simultaneously in order to fully utilize advanced IT systems, and draw more business performance with the IT systems.

5.2 RQ2: Comparison with BABOK Tasks

To evaluate the proposed DBA method, we compared the activities of DBA method with the tasks of BABOK [4]. BABOK is structured in two layers of *knowledge areas* and *tasks* in each knowledge area. From the authors' viewpoint, the BABOK is *process oriented* in the sense that it describes the workflow of the tasks to analyze the business analysis. In BABOK, there are substantially five knowledge areas of business analysis, and 25 tasks in total. Looking at the tasks of BABOK, it is comparable to the activities of the proposed DBA process. The authors compared 34 activities of the DBA process and 25 tasks of BABOK, and found the following characteristics.

(1) Concept of Business Analysis and Structure of Analysis Process
BABOK focuses on the analysis of business process by its very definition, and pays less attention to IT since IT is considered as a part of the solution. On the other hand, DBA process is based on the concept of digital business, and analyze both business system and IT system simultaneously. This is a fundamental difference between BABOK and DBA process. From the authors' viewpoint, analysis of business process is inseparable from that of IT systems. Therefore, it is more preferable to use the DBA process in the context of digital business analysis.

(2) Granularity of Task/Activity
The granularity of BABOK tasks is rather coarse. Therefore, in order to apply the BABOK, it might be necessary to compensate some additional guide. On the other hand, the granularity of the process and activities of DBA process is relatively small. In the development of DBA process, the authors conducted intensive discussions with many practitioners on the granularity and boundary of processes and activities. From the experience, the authors believe it is hard to identify a unique process model which can meet all types of business analysis by any stakeholders and in any domains. Therefore, the proposed DBA process is a reasonable compromise at a reasonable abstraction from the discussions.

6 Conclusions

This article proposes a DBA (Digital Business Analysis) method. The concept underlying the proposed method is the digital business systems, in which business systems are inseparable from information systems. Therefore, the analysis of business systems and information systems is also inseparable, and should be conduct simultaneously. Therefore, the DBA process, the core technique of DBA method, embraces the characteristics of digital business systems. The proposed DBA method also emphasizes the collaboration between the business departments and IT department, and specifies the roles of two department in the DBA method.

By the comparison of the proposed DBA process with real business analysis processes conducted in Japanese companies, and comparison with BABOK proved the feasibility of the proposed DBA process.

As the future work, the authors further study the structure of the DBA process, and extend the process to include management and monitoring capabilities. The authors also plan a series of feasibility studies with real applications.

Acknowledgement. The authors acknowledge the constructive discussions of the members of BRE (Business Requirements Engineering) research group.

References

1. Aoyama, M., et al.: A model and architecture of REBOK (Requirements Engineering Body Of Knowledge) and its evaluation. In: Proceedings of APSEC 2010, IEEE Compute Society, pp. 50–59, November–December 2010
2. Aoyama, M.: Bridging the requirements engineering and business analysis toward a unified knowledge framework. In: Link, S., Trujillo, J.C. (eds.) Proceedings of the MReBA 2016, ER 2016 Workshops, LNCS, vol. 9975, pp. 146–160. Springer, Cham (2016)
3. Benedict, T., et al.: BPM CBOK Version 3.0: Guide to the Business Process Management Common Body of Knowledge, ABPMP International/Createspace (2013)
4. IIBA: A Guide to the Business Analysis Body of Knowledge (BABOK Guide), Version 3.0, IIBA (2015)
5. ISO/IEC/IEEE 29148:2011 Software and Systems Engineering - Life Cycle Processes, Requirements Engineering, ISO (2011)
6. Jeston, J., Nelis, J.: Business Process Management, 3rd edn. Routledge, London (2014)
7. Jisa Rebok, W.G. (ed.) Requirements Engineering Body Of Knowledge (REBOK), Version 1.0, Kindaikagakusha (2011). (In Japanese). http://www.re-bok.org/en/
8. Paul, D., et al.: Business Analysis, 2nd edn. British Informatics Society, Swindon (2010)
9. Ullah, A., Lai, R.: Modeling business goal for business/IT alignment using requirements engineering. J. Comput. Inf. Syst. **51**(3), 21–28 (2011)
10. Wang, R.: Disruptive Digital Business. Harvard Business Review Press, Boston (2015)

Integrated Framework for Software Requirements Analysis and Its Support Tool

Andre Rusli(✉) and Osamu Shigo(✉)

Graduate School of Information Environment,
Tokyo Denki University, Chiba, Japan
andreruslil9@gmail.com, shigo@mail.dendai.ac.jp

Abstract. Our research focuses on integrating four popular methods which are i* framework, KAOS model, Problem Frames, and Message Sequence Chart, and suggests a new integrated framework for requirements analysis. We also implement the framework on a case study to prove its usability. Moreover, our research aims to develop a support tool to help users in using the proposed integrated framework that will assist user in drawing diagrams and enable diagram auto-generation and inconsistency prevention.

Keywords: Requirements analysis · Goal models · i* framework · Problem frames · KAOS · Message sequence chart · Integrated framework · Support tool

1 Introduction

Broadly speaking, software systems requirements engineering (RE) is the process of discovering that purpose, by identifying stakeholders and their needs, and documenting these in a form that is amenable to analysis, communication, and subsequent implementation [8]. The importance of this process arises from the fact that it is done in the early phase of software development. Requirements engineering [1, 6] provides the base foundation of the software system that is being developed, hence, it is fundamental to prioritize this process before doing other engineering activities such as design or even to start programming. Our research focuses on four methods which we considered can be integrated into one framework [9] so that they can complete each other's advantages and disadvantages, i* framework, KAOS model, Message Sequence Chart, and Problem Frame.

Goal based models support requirements engineers to elicit the objectives of the system development, however goals are hierarchical, hence, it sometimes becomes difficult to determine where a goal is situated in the hierarchy and how it relates to the problem context [7]. At the same time, describing goals and its relationship with the stakeholder in the system is still a fundamental work in RE. I* framework focuses more on the early phase of requirement engineering by describing the dependency relationships between stakeholders and their internal intentional relationships. However, the models lack in describing the timely order of the elements in the system. Furthermore, KAOS goal model [3, 4] supports risk analysis regarding the system goals which is critical in RE. Problem frames [5], on the contrary, are focused on modeling

S.-W. Lee and T. Nakatani (Eds.): APRES 2016, CCIS 671, pp. 132–140, 2016.
DOI: 10.1007/978-981-10-3256-1_10

and understanding the context of requirements engineering in the real world. However, they lack describing the goals of the system and does not support risk analysis.

One more method that are considered in our research is Message Sequence Chart (MSC) [2]. MSC supports the analysis of dynamic behaviors of the system. By analyzing the dynamic behaviors, it helps the stakeholders and engineers to better understand the system. However, MSC lacks in analyzing the system goals and problem context. The integration of these four methods is expected to able to provide a framework which can support the elicitations of early-phase requirements that analyze goals and problem context, while also support the analysis of the dynamic behavior of current and future system and describe risks and their possible solutions.

We found that integrating these methods have some shortcomings too. Manually draw the whole diagram is quite troublesome. While there are some elements that keep appearing in several diagrams that can cause redundancy of work, there are another concerns of the probability of inconsistency between each diagrams if they are manually drawn by hands. To solve these problems, we decided to develop a support tool for the integrated framework. This support tool will assist the users in implementing our framework to analyze requirements.

2 Research Question

The research that we are conducting is aimed to answer the following research questions:

1. **RQ1**. *How is the proposed integrated framework useful for requirements analysis?*
2. **RQ2**. *How can the support tool being developed help user in using the proposed integrated framework?*

The answer to RQ1 will show that our work contributes positively to help engineers analyze requirements in software development. While RQ2's answer explains that a support tool is needed to assist user in implementing the proposed framework. These answers are explained in Sects. 4, 5, and 6, and will be concluded in the conclusion of this paper.

3 Integrated Framework for Software Requirements Analysis

Our framework utilizes i*'s concept of seeing the stakeholders as actors having intentional properties such as goals, beliefs, abilities and commitments [10], then describes the dependency relationship between them. Moreover, to improve the dependency diagram which concept was adapted from i* framework, we added Problem Frames (PF)'s requirement constraints concept into the dependency diagram. PF method believes that it is important to look away for a moment from the computer world and focus to the real world. By focusing more at the real world which the software system will be built on, it is easier to extract constraints that exist in the real world and apply them into the analysis that we already built on the dependency diagram.

Those relations with in the dependency diagram shows goal, resource, and/or task dependency between actors. However, it is also essential to know what each actor is supposed to do within themselves, thus the next step of our integrated framework is to analyze the hierarchy inside each actors, adapted from i*'s strategic rationale model. The previous diagrams shows a detailed description on how actors depend on each other, whether it is goal, resource, or task dependencies. However, the dynamic behavior of the system is not described, yet it is quite essential to know which resource goes first and which goes later. Our framework utilizes Message Sequence Chart (MSC) to analyze these dynamic behaviors. The other fundamental process in requirement engineering is risk analysis. Our framework imported the concept from KAOS' obstacle analysis to do this. Basically, we pick task(s) that we think is risky from the previous diagrams and analyze the risks that might happen in the future and describe the possible solutions regarding those risks.

4 Case Study Implementation

Our work implemented the integrated framework to analyze requirements of a simplified case study that is taken from the research by Cailiau, Damas, Lambeau, Lamsweerde (2013), the Barbados Car Crash Management System (bCMS) [3], to test whether the framework is useful for requirement analysis. The main objective is to develop a system which is intended to coordinate the communication between a fire station coordinator (FSC) and a police station coordinator (PSC) to handle crises, which is a car crash accident, in a timely manner.

Figure 1 shows the first diagram in the framework that needs to be drawn in the first place, which is the dependency diagram. Four actors are recognized in the system,

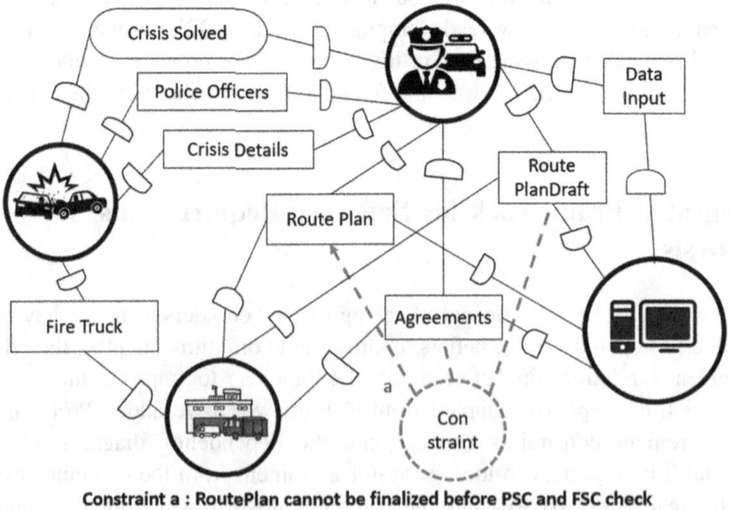

Fig. 1. Dependency diagram

Victim, Police Station Coordinator (PSC), Fire Station Coordinator (FSC), and System, which is the computer system that will be developed in the project. The diagram in Fig. 1 is focused on describing the dependency relationships between those four actors. Moreover, the direction of half-circles on lines that connect actors shows the dependency relations between actors. This concept is adapted from i* framework's strategic dependency diagram.

As explained earlier, the dependency diagram in Fig. 1 uses i*'s notation on pointing which actor depends on which actor on which resource. For example, Computer System depends on PSC and FSC to provide agreement, while PSC and FSC depend on System to provide route plan. Moreover, requirement constraint, which is drawn as a dotted blue circle, constraints the resource "Route Plan" in such that a Route Plan must be checked and agreed by both PSC and FSC to be finalized. This concept is taken from Problem Frames' concept on describing the requirement exists in the real world, apart from the computer. This part is what makes our dependency diagram a little bit different from i*'s strategic dependency diagram. Requirement constraint will later affect the making of message sequence chart for describing the scenario in the system.

Figure 2 shows one example of an actor, the PSC actor, and its hierarchy. PSC's main goal is to handle crisis and to do so it needs has three tasks, which include sending police car, checking route plan draft, and inputting data to the system. The next step, which is analyzing dynamic behaviors, is shown in Fig. 3 and done using Message Sequence Chart, taking variables such as actors, resources, and constraint from the previous dependency diagram.

Fig. 2. PSC (Police Station Coordinator) actor's goal and task hierarchy

The next step of our framework is to analyze obstacle that might exists in the sys-tem and possible solutions for each of those obstacles/risks. Figure 4 shows the risk analysis diagram of the "Send Police Vehicle" task from bCMS case study. It can be seen that there might be a risk of traffic jam when police cars are sent into the scene. Regarding this risk, there are two possible solutions. These solutions are describe in the risk analysis diagram, then it is up to the stakeholder to decide whether the solutions should be applied into the system or not.

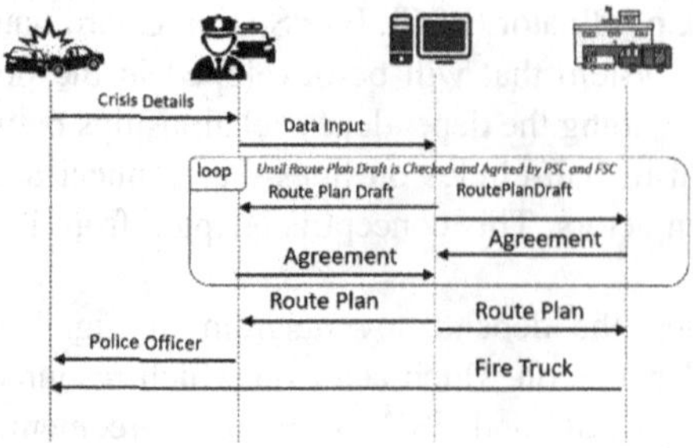

Fig. 3. Message sequence chart

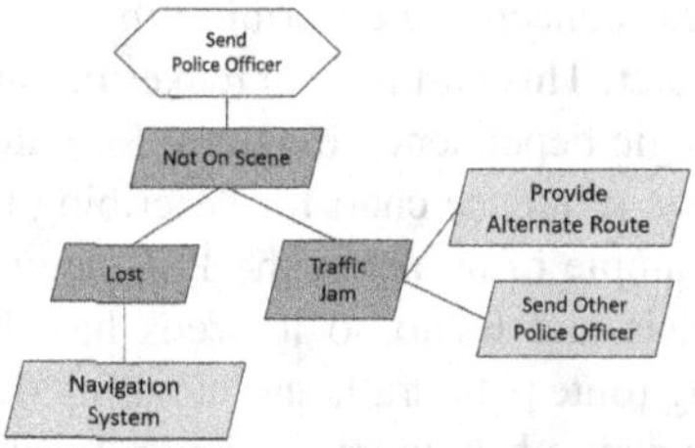

Fig. 4. Risk analysis of "SendPoliceCar" task

5 Support Tool

The next step of our research is to tackle the difficulties that are found in implementing the proposed integrated framework to analyze requirements. Those difficulties include possibility of inconsistency and redundancy and increased complexity in drawing the diagrams in the framework. One of the critical thing in this framework is that the name of components must be always constant because it will not just used in one diagram only, but in other related diagram. Naming mistakes may cause confusions in analyzing the whole requirements. Inconsistency in this context is possible to occur when the diagrams in the framework are all written manually by hand. While redundancy is caused mainly of the same reason that there will be same components that keep appearing in not only one diagram, which means the user needs to keep on drawing the same components in different diagrams.

In order to overcome those disadvantages, our research also develops the support tool to make it easier for engineers to use our framework and to help user to get better understanding of the diagrams. Our support tool offers help to draw diagrams with computer on a web application, so it is no necessary for engineers to hand-write all of the diagrams which means that it can reduce work and the efforts needed. Figure 5 shows the main interface of our support tool to draw the dependency diagram.

Moreover, actors on the diagram can be enlarged by double clicking them and user will be able to edit or view the goal and task hierarchy inside each actor, as shown in Fig. 6. The resources drawn outside the actor's circle are representing the resources that it depends on other actors or that are depended on them by other actors.

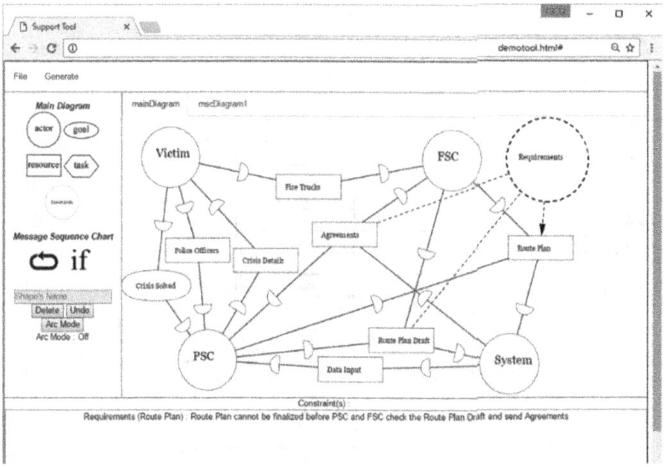

Fig. 5. Dependency diagram creator tool

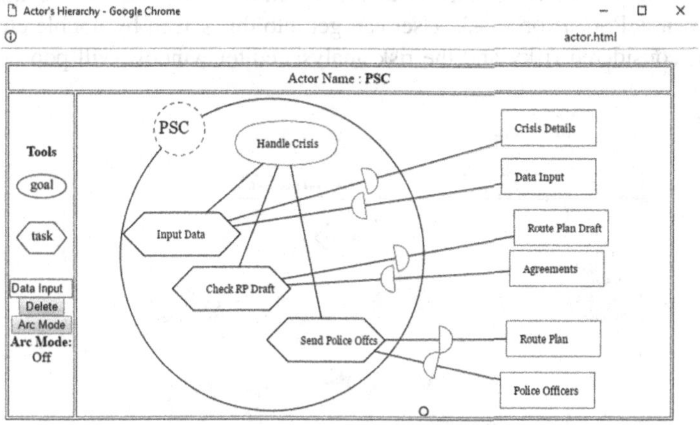

Fig. 6. Actor's hierarchy editor

Tasks inside and outside each actor can be marked as risky by double-clicking them for further analysis in the later process of the framework.

The proposed support tool will also enable diagram auto-generation from dependency diagram into MSC. This is possible because actors and resources presented in the dependency diagram can be reused in MSC to explain their behavior. First, the tasks and actors from dependency diagram are imported just as they are ordered in the dependency diagram. After they are generated, user can re-arrange the timely order of the tasks by clicking and dragging the resources drawn on the diagram, so the ideal scenario of the system can be understood, which is critical in analyzing requirements, as shown in Fig. 7.

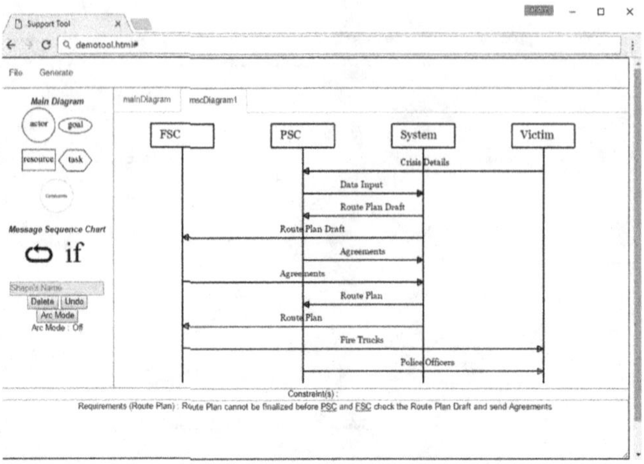

Fig. 7. Message sequence chart editor

Figure 8 shows the risk analysis diagram editor for the task of sending police officer, which is PSC actor's task. User can get into this screen by double-clicking the task that is considered risky and the risk analysis editor window will pop-up.

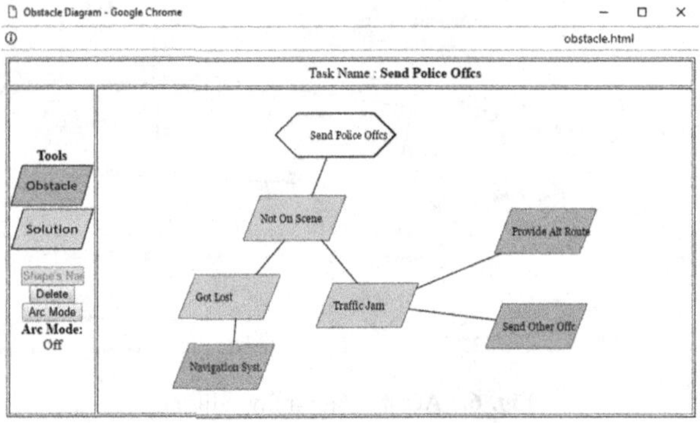

Fig. 8. Risk diagram editor

Using this support tool lowers the effort needed from the user to draw the diagrams, enables auto-generation to prevent inconsistency, and reduce the possibility of redundant work. Diagrams can also be saved and loaded as wished by the user, thus, provide assistance to keep track of the activities that has been done.

6 Conclusion and Future Work

Describing dependency relationship between actors in the system, hierarchies inside actors, dynamic behaviors, and risk analysis are the activities involved in our proposed integrated framework for requirements analysis in the early phase of software development. We implemented and showed the usability of the proposed framework on a case study of Barbados Car Crash Management System and as a result, compared to using one requirements analysis method only, which limits our view upon the whole requirements, combining the advantages of i* framework, KAOS model, MSC, and Problem Frames enables us to complete each method's disadvantages and provide better requirements description, which answers to *RQ1*.

The difficulties we found on using this framework, which includes complicated views, possibilities of inconsistency, and redundancies, are tackled by the support tool that we have been developing. The support tool enables user to simply draw diagrams and enables diagram auto-generation which helps prevent inconsistency and redundancy of work, thus, compared to manual drawing, assists users to easily implements the integrated framework in requirements analysis, answering *RQ2*.

The future of our work will include the completion of the support tool, mainly because the current tool is still in its prototype version. After the support tool is completed, it needs to be tested and evaluated by having users to actually use the tool to analyze requirements and get their critics and comments to better improve the build of support tool software. Moreover, it is also interesting to have real industrial people to implement our framework, so not only from case study, it can be evaluated in the actual world too.

References

1. Bray, I.K.: An Introduction To Requirement Engineering. Addison Wesley, Reading (2002)
2. Broy, M.: The essence of message sequence chart. In: Proceedings of the International Symposium on Multimedia Software Engineering, pp. 42–47. IEEE (2000)
3. Cailliau, A., et al.: Modeling car crash management with KAOS. In: Proceedings of the 3rd International Comparing Requirements Modeling Approaches (CMA@RE), pp. 19–24. IEEE (2013)
4. Cailliau, A., et al.: Modeling Car Crash Management with KAOS, UCL (2013). kaos.info.ucl.ac.be/bcms.html
5. Jackson, M.: Problem Frames: Analysing and Structuring Software Development Problems. Pearson Education, Harlow (2001)
6. van Lamsweerde, A.: Requirement Engineering: From System Goals to UML Models to Software Specifications. John Wiley & Sons, Chichester (2009)
7. Mohammadi, N.G., Alebrahim, A., Weyer, T., Heisel, M., Pohl, K.: A framework for combining problem frames and goal models to support context analysis during requirements engineering. In: Cuzzocrea, A., Kittl, C., Simos, D.E., Weippl, E., Xu, L. (eds.) CD-ARES 2013. LNCS, vol. 8127, pp. 272–288. Springer, Heidelberg (2013). doi:10.1007/978-3-642-40511-2_19

8. Nuseibeh, B., Easterbrook, S.: Requirements engineering: a roadmap. In: Proceedings of Conference on the Future of Software Engineering, pp. 35–46. ACM (2000)
9. Rusli, A., Shigo, O.: Integrated framework for software requirement analysis. In: REFSQ Workshop (2016). http://ceur-ws.org/Vol-1564/paper11.pdf
10. Yu, E.: Towards modelling and reasoning support for early-phase requirement engineering. In: Proceedings of the Third IEEE International Symposium on Requirement Engineering, pp. 226–235 (1997)

Author Index

Printed in the United States
By Bookmasters